# *New Results*

* *Unification of QED, Weak Interactions, Strong Interactions and Quantum Gravity.*

* *General Formulation of Divergence-free Quantum Field Theories. Detailed discussions of unitarity and special relativity showing these theories are physically acceptable.*

* *Divergence-free massive vector bosons quantum field theory: No need for the Higgs mechanism.*

* *"Low Energy" Limit of elementary particle sector of unified theory approximates the Standard Model (& QED) to extreme accuracy.*

* *Suggests possible doubly charged dilepton, and other exotic, resonances.*

* *The "Large Distance", classical limit of Quantum Gravity sector is General Relativity.*

* *No ultra-light Black Holes. Gravity is repulsive (anti-gravity) at ultra-short distances.*

* *Two-tier gravity "saves" the concept of a space-time point by evading Wigner's argument against it.*

* *Based on experimental data a preferred local reference frame defined by Cosmic Background Radiation is shown to exist in each locale. Preferred local inertial frames are used in two-tier quantum gravity whose dynamical equations are invariant under general relativistic transformations but whose "ground state" breaks the invariance down to invariance under special relativity.*

* *A New form of hidden dimensions: Quantum Dimensions – dimensions implemented via a quantum gauge field.*

* *A New method in the Calculus of Variations – composition of extrema.*

# *Quantum Theory of the Third Kind*

## *A New Type of Divergence-free Quantum Field Theory Supporting a Unified Standard Model of Elementary Particles and Quantum Gravity based on a New Method in the Calculus of Variations*

*Some Other Books By Stephen Blaha*

*The Equivalence of Elementary Particle Theories and Computer Languages: Quantum Computers, Turing Machines, Standard Model, Superstring Theory, and a Proof that Gödel's Theorem Implies Nature Must Be Quantum* (ISBN: 0974695823, Pingree-Hill Publishing, Auburn, NH, 2005)

*Quantum Big Bang Cosmology: Complex Space-time General Relativity, Quantum Coordinates™ Dodecahedral Universe, Inflation, and New Spin 0, ½, 1 & 2 Tachyons & Imagyons™* (ISBN: 0974695815, Pingree-Hill Publishing, Auburn, NH, 2004)

*A Finite Unified Quantum Field Theory of the Elementary Particle Standard Model and Quantum Gravity Based on New Quantum Dimensions™ and a New Paradigm in the Calculus of Variations* (ISBN: 0972079599, Pingree-Hill Publishing, Auburn, NH, 2003)

*Cosmos and Consciousness Second Edition* (ISBN: 0972079548, Pingree-Hill Publishing, Auburn, NH, 2002)

*The Life Cycle of Civilizations* (ISBN: 0972079580, Pingree-Hill Publishing, Auburn, NH, 2002)

*A Scientific Inquiry into the Nature of God, the Spiritual, and Near Death Experiences* (ISBN: 0972079521, Pingree-Hill Publishing, Auburn, NH, 2002)

*A Java™ Programming Introductory and Intermediate Course Second Edition* (ISBN: 097207953X, Pingree-Hill Publishing, Auburn, NH, 2002)

*C++ for Professional Programming With PC and Unix Applications* (ISBN: 1850328013, International Thomson Publishing, Boston, 1995)

*The Reluctant Prophets: Has Science Found God?* (ISBN: 0759663041, 1stBooks Library, Bloomington, IN, 2001)

*The Rhythms of History: A Universal Theory of Civilizations* (ISBN: 0972079572, Pingree-Hill Publishing, Auburn, NH, 2002)

Available on Amazon.com, bn.com and other Internet sites as well as better bookstores (from Ingram distributors).

**Cover Credits**

# Quantum Theory of the Third Kind

## A New Type of Divergence-free Quantum Field Theory Supporting a Unified Standard Model of Elementary Particles and Quantum Gravity based on a New Method in the Calculus of Variations

Stephen Blaha, Ph.D.*

Pingree–Hill Publishing

* sblaha777@yahoo.com

Pingree-Hill Publishing
P. O. Box 368
Auburn, NH 03032 USA

Or email to: baliltd@compuserve.com

ISBN: 0-9746958-3-1

This book is printed on acid free paper.



rev. 00/00/01

# PREFACE TO THE SECOND EDITION

This is the second edition of *A Finite Unified Quantum Field Theory of the Elementary Particle Standard Model and Quantum Gravity: Based on New Quantum Dimensions™ & a New Paradigm in the Calculus of Variations*. It contains new material such as a proof that two-tier quantum field theories are relativistic and corrects some typos in the first edition. It also provides a deeper discussion of the basis and rationale of this new form of quantum theory.

# PREFACE

The Standard Model of Elementary Particles enjoys a measure of success in accounting for experimental results that is somewhat amazing in view of its "hodge-podge" nature, and in view of the manner in which it evolved from the 1940's until the present. Some issues that remain to be explicated are CP violation; the origin, and rationale, of the internal symmetries; the existence and origin of Higgs particles; and the unification of the Standard Model with Quantum Gravity. Practically the Standard Model is the only viable theory for performing calculations to compare to experiment.

The quantum field theoretic aspects of the Standard Model have been hitherto considered as satisfactory because the divergences that appear within it have been brought under control through renormalization techniques so that meaningful calculations can be made and compared with experiment. However these infinities (divergences) remain a source of uneasiness—particularly since known renormalization techniques cannot handle the infinities that crop up in Quantum Gravity—thus precluding a unified theory of all interactions.

One major attempt to solve the divergence issue as well as create a unified theory is Superstring theory. This theoretic approach may eventually be successful but the absence of any experimental evidence for the plethora of particles that superstring theories predict raises major questions as to its physical reality. Nature seems to be more simple than Superstring theory would have us believe. The use of compactified dimensions to generate internal symmetries also seems to beg the question. Compacted dimensions are little more than an artifice to insert extra dimensions. The physics of the compactification is unclear.

The type of quantum field theory theory presented *in this book* assumes that the universe has additional quantum dimensions that serve to eliminate divergences in quantum field theory – every particle has a quantum fluctuating cloud of quanta generating quantum dimensions around it. The dimensions are implemented as quantum fields and thus have no need for compactification.

The new type of theory, two-tier quantum field theory, is based on a new paradigm for the Calculus of Variations described in Appendix A. It leads to a new approach to quantization that we call quantum theory of the third kind. First there was quantum mechanics. Then there was quantum field theory. The new kind of quantum field theory defines quantum coordinates as the variable of quantum fields. (In conventional quantum field theory the coordinates are c-numbers – not operators.)

*To my Wife, Margaret*

*With Love*

# CONTENTS

# LIST OF FIGURES

# 1. Quantum Dimensions vs. Classical Dimensions

*All beginnings are obscure.*
*H. Weyl – Space, Time, Matter*

## Beyond 4-Dimensional Space-time

There have been countless attempts since the 1920's to use additional dimensions beyond the known four dimensions of space and time to explicate and unify the fundamental forces of nature. The most noteworthy *recent* attempts along these lines have been Superstring theories and Technicolor theories.

In the opinion of this author the efforts in these directions are not justified by the results. The physics that these theories attempt to describe is simpler than the formulation of the theories with much fewer particles and interactions. Assigning high masses to undiscovered particles and placing undiscovered forces in the high energy regime beyond the limits of accelerators does not seem to be a satisfactory approach. Extrapolations of theories in the past, without the guidance and confirmation of experiment, have usually not been successful.

Therefore it seems reasonable to develop a deeper, sounder formulation of the Standard Model *as it is now* since it was developed through a close interplay of theory and experiment. It remains the preeminent theory of elementary particles—actually the *only* experimentally acceptable theory of elementary particles.

This book attempts to establish a deeper framework for the Standard Model that enables it to be combined with quantum gravity to form a divergence-free unified quantum field theory of nature.

It is clear from the existence of internal symmetries in the Standard Model that something is "going on" inside particles which is outside the framework of normal space and time. Otherwise, there would be no internal symmetries.

Therefore it is reasonable to consider the possibility of extra dimensions beyond normal space-time. The open question is how these extra dimensions enter into physical theory. Superstring theory appears to go too far in terms of the numbers of dimensions and the particles that it requires – not to mention – the complexity of its mathematics, appears to preclude all but the simplest calculations. Technicolor and extended Technicolor theories also introduce substantial additional complexities in order to explicate the pattern of symmetries of the Standard Model. We face the question of whether the cure is worst than the disease in these approaches.

Thus we ask if a more tractable theory is possible. Our first requirement is that it would improve the Standard Model by taming the divergences in quantum field theory so that the Standard Model can be unified with Quantum Gravity. This author has suggested[1] an alternate form of quantum field theory, two-tier quantum field theory, that is ultra-violet divergence-free to all orders for theories of the type of the Standard Model, and for Quantum Gravity. This theory not only resolves divergence issues in the Standard Model and Quantum Gravity but also eliminates the singularities at the point of the Big Bang in Cosmology.[2]

This theory is *not* based on extra dimensions that become compactified as in Superstring theories. Rather it postulates that extra dimensions are directly generated by a free quantum field and thus constitute *quantum dimensions*™. Quantum dimensions are fluctuating quantum degrees of freedom that make each elementary particle a "fuzzy ball" that partly exists in imaginary space as well as real space.

*Ideally, in this author's view, we would start with a concept of a pre-dimensional entity – an entity that is perhaps initially formless which evolves, perhaps through a form of self-organization, to develop dimensions, energy and quantum particles. Physics is far from developing such a grand scheme and must, at present, content itself with assuming the existence of dimensions, a Lagrangian of some sort, and quantum dynamical entities (particles). Thus we will make assumptions as to the number and nature of additional dimensions.*

## Quantum Dimensions vs. Classical Dimensions

We are all familiar with the concept of dimension. Euclid gave a geometrical definition of dimensions and a procedure for determining the number of dimensions: move in a straight line for a distance; then make a 90° turn; again move in a straight line in the new direction for a while; then make a 90° turn such that the new direction is perpendicular to the previous two directions of motion; continue this procedure until

---

[1] See Blaha (2003) – the first edition of this book.
[2] See Blaha (2004).

it is no longer possible to make a 90° turn to move in a new direction. This process establishes the direction of dimensions and their number in flat space.

From a Cartesian point of view the number of dimensions can initially be simply viewed as the number of independent coordinate axes. Each axis is broken into intervals in such a way as to allow us to specify a position in space with an ordered set of numbers. We can then define functions that depend on these ordered sets of numbers – the coordinates of a point. Such a function can then have a range of values as the coordinates change from point to point in space. We can denote this function as $f(x_1, x_2, x_3, \ldots x_n)$ if the space has n dimensions.

The range of the dimensions in a flat Cartesian space is usually from $-\infty$ to $+\infty$. One can also specify cyclic or compact dimensions that "form a circle" – a coordinate can range in value say from 0 to $2\pi$. The point at $2\pi$ can then be made to coincide with the point at 0 so that one can view the dimension as a circle.

Another approach, that was first introduced by Blaha(2003), is to use a new type of coordinate – a quantum coordinate. To understand this concept we imagine a 3-dimensional space of the normal sort with coordinates: x, y, z and values ranging from $-\infty$ to $+\infty$. Now suppose, for example, we introduce a sine function:

$$G(x, y) = \sin(x + y) \tag{1.1}$$

and require

$$z = G(x, y) \tag{1.2}$$

Then a function of the three coordinates of space f(x, y, z) becomes

$$f(x, y, z) = f(x, y, G(x, y)) \tag{1.3}$$

on the surface defined by eq. 1.2. We see f still has values at each point in 3-dimensional space. However, since eq. 1.2.2 defines a 2-dimensional surface within the 3-space the expression for f in eq. 1.3 is properly viewed as defined on that surface.

Now if we replace G(x, y) with a second quantized field Q(x, y) in eq. 1.3 then we obtain a qualitatively new entity:

$$f(x, y, Q(x, y)) \tag{1.4}$$

is now an operator expression. (We will assume that infinities and other issues are not present or resolvable.) Eq. 1.4 in itself does not have a value. It obtains a value when evaluated for quantum states $|q>$

$$q(x, y) = \langle q | f(x, y, Q(x, y)) | q \rangle \quad (1.5)$$

Thus the replacement of the z coordinate in f(x, y, z) with a field operator gives us a qualitatively new entity with well-defined values only when evaluated between states.

In a sense f(x, y, Q(x, y)) is dependent on three unknown quantities – the values of x and y, and the expectation value of f(x, y, Q(x, y)) between (as yet unidentified) quantum states. Thus we can regard f(x, y, Q(x, y)) as depending on two coordinates x and y, and on a quantum coordinate™ whose value is an undetermined quantum fluctuation quantity that only becomes known upon taking an expectation value.

We thus have developed a new form of dimension that is quite properly called a quantum dimension with values that are not c-numbers but in fact are q-numbers (determined only by taking expectation values between quantum states).

*Quantum Theory of the Third Kind*

First there was quantum mechanics – developed essentially in the period from 1914 - 1926. Quantum Mechanics postulated that position and momentum were non-commuting operators and so the position and momentum of a particle could not be measured simultaneously to arbitrary accuracy. Instead they satisfied the Heisenberg Uncertainty Principle:

$$\Delta x \Delta p \geq \hbar \quad (1.6)$$

Then quantum field theory (second quantization) was developed roughly in the period from 1935 – 1960. Quantum theory postulated that any quantum field and its conjugate momentum operator satisfied an uncertainty principle:

$$\Delta\phi \Delta p_\phi \geq c_\phi \hbar \quad (1.7)$$

for any quantum field $\phi$ and its conjugate momentum $p_\phi$ where $c_\phi$ is a factor dependent on the nature of the quantum field $\phi$. As a result the arguments of the field operators $\phi$ - the position coordinates – were treated as ordinary parameters – c-numbers. Eq. 1.6 and eq. 1.7 were shown[3] to both be required for a consistent quantum theory. The standard example considered the effect of using a quantum electromagnetic field to measure the position and location of a particle.

[3] Heitler (1954) pp. 79-86.

We will suggest that there is a further quantum formulation beyond first first and second quantization in which quantum fields are functions of quantum coordinates. This type of *quantum theory of the third kind* resolves the divergence problems that have plagued second quantization.

Loosely speaking by making the coordinate arguments of quantum fields “fuzzy” we avoid the infinities associated with evaluating products of quantum field operators at precisely the same point. We will proceed to consider quantum field operators that are functions of quantum coordinates in the remainder of this book.

## Quantum Coordinates in 4-dimensional Space-time

Now that we see the nature of a quantum dimension we will turn to developing the form of space-time for a universe with both ordinary space-time dimensions and additional dimensions to resolve divergence problems. In Blaha (2003) and Blaha (2004) we were concerned with resolving the divergences of the Standard Model quantum field theory and the divergences of Quantum Gravity as well as the singularity issue at the point of the Big Bang in the Standard Model of Cosmology. We pointed out all these issues could be resolved by using complex quantum coordinates

$$X_\mu(y) = y_\mu + i\, Y_\mu(y)/M_c^{\,2} \qquad (1.8)$$

where $Y^\mu(y)$ is a real quantum field with properties identical to the free electromagnetic quantum field of Quantum Electrodynamics, $y^\mu$ is a Minkowski space-time 4-vector, and $M_c$ is a large mass that is presumably of the order of or equal to the Planck mass.

All particle (boson and fermion) quantum fields were defined as q-number functions of the quantum coordinates $Y^\mu(y)$ and not of $y^\mu$ directly. Using a new method in the Calculus of Variations (Appendix A) that we called the “composition of extrema” we developed quantum field theories – called *two-tier quantum field theories* – for particles of spin 0, ½, 1, and 2.

For example, the two-tier Lagrangian for a free scalar particle is

$$I = \int \mathscr{L}_S\, d^4y \qquad (1.9)$$

with

$$\mathscr{L}_S = J\, \mathscr{L}_F(\phi(X), \partial\phi/\partial X^\mu) + \mathscr{L}_C(X^\mu(y), \partial X^\mu(y)/\partial y^\nu) \qquad (1.10)$$

where $\phi$ is a scalar field, J is the Jacobian for the transformation from $X^\mu(y)$ coordinates to $y^\mu$ coordinates, and

$$\mathscr{L}_F = \tfrac{1}{2}\,[(\partial\phi/\partial X^\nu)^2 - m^2\phi^2] \tag{1.11}$$

If we define $X^\mu(y)$ using eq. 1.8 then

$$\mathscr{L}_C = -\tfrac{1}{4}\, F_Y^{\ \mu\nu} F_{Y\mu\nu} \tag{1.12}$$

where

$$F_{Y\mu\nu} = \partial Y_\mu/\partial y^\nu - \partial Y_\nu/\partial y^\mu \tag{1.13}$$

Upon variation in $\phi$ with $y^\mu$ (and thus $Y^\mu(y)$) held fixed (see Appendix A for a discussion of composition of extrema in the Calculus of Variations) we find

$$\partial\mathscr{L}/\,\partial\phi\ - \partial/\partial X^\mu\ [\partial\mathscr{L}/\partial(\partial\phi/\partial X^\mu)] = 0 \tag{1.14}$$

which gives us the Klein-Gordon field equation for $\phi$

$$(\Box + m^2)\ \phi(X) = 0 \tag{1.15}$$

where

$$\Box = \partial/\partial X^\nu\, \partial/\partial X_\nu \tag{1.16}$$

A Fourier representation of the solution of eq. 1.15 is:

$$\phi(X) = \int dp\ \delta(p^2 - m^2)\theta(p^0)\ [a(p){:}e^{-ip\cdot X}{:} + a^\dagger(p){:}e^{ip\cdot X}{:}] \tag{1.17}$$

where a(p) is a function of p, $^\dagger$ indicates hermitean conjugation, and : : indicate normal ordering of the q-number expression in $X^\mu(y)$.

The manifold defined by $X^\mu(y)$ is found by variation of the lagrangian with respect to $Y^\mu(y)$. In brief it yields field equations (and gauge invariance) that are identical to the case of the electromagnetic field. This scalar particle case and other cases are considered in detail in the following chapters. So we will defer further consideration of the scalar particle case until then.

The following points were proved in Blaha (2003) and Blaha (2004):

1. Free field theories created with the two-tier formulation (including quantum gravity) have propagators that are the same as those in the corresponding conventional quantum field theory except that the Fourier representation of each particle propagator contains a gaussian momentum factor. This gaussian factor eliminates all ultra-violet divergences in perturbative quantum field theories such as the Standard Model and Quantum Gravity. The gauge invariance of $Y^{\mu}(y)$ was required in order to have well-behaved gaussian factors throughout the momentum region. Blaha (2003).

2. The resultant theories were proven to satisfy unitarity. Blaha (2003).

3. The resultant theories were proven to be Lorentz invariant in any gauge of $Y^{\mu}(y)$. Blaha (2004).

Thus this new approach to quantum field theory (two-tier quantum field theory), when applied to the Standard Model and Quantum Gravity, results in divergence-free quantum field theories enabling us to create a unified, divergence-free quantum field theory of all the forces of nature. (The question of creating this type of quantum field theory in curved space-time was successfully addressed in Blaha (2004).)

The nature of the point of the Big Bang in Cosmology has been an ongoing issue. When two-tier quantum field theory is applied to the Big Bang we find that it suggests the universe is very dense region of finite size and temperature at the time of the Big Bang with an inhomogeneous generalized Robertson-Walker metric. The Einstein equations of this metric are separable: one equation is dominated by classical physics; the other equation is quantum in nature. The universe is shown to rapidly change into a standard Robertson-Walker universe with the usual scale factor. Thus the Standard Model of Cosmology emerges shortly after the Big Bang but the singularity of the Standard Cosmological Model at the Big Bang is eliminated Blaha (2004).

Thus the two-tier quantum field theoretic approach eliminates the divergence problems of the Standard Model, Quantum Gravity and the Standard Cosmological Model. In addition it justifies treating the Big Bang epoch of the universe as having an ultradense, quasi-free energy density in the form of a perfect fluid.[4] This result follows from the factorization of the scale factor of the generalized Robertson-Walker metric into a factor that is dominated by the macroscopic energy density and is thus primarily

---

[4] A recent experiment in which gold nucei collided at high energy confirms that the collision region briefly contained a perfect fluid consisting of a quark-gluon plasma. This result is consistent with our two-tier Standard Model and the two-tier cosmological model developed in Blaha (2004).

classical in nature; and another factor embodying quantum effects that is independent of the energy density.

## Features of Two-Tier Quantum field Theories

The currently known theories of fundamental interactions and elementary particles fall into two broad categories: conventional quantum field theories and string-based theories that are united with supersymmetry to form superstring theories. Many physicists believe that only theories of these types can meet the reasonable physical requirements of Lorentz invariance and unitarity (positive probabilities summing to unity) while accounting for spin and internal quantum numbers.

It appears that our new form of elementary particle theory is viable. In this type of theory quantum fields are functions of q-number coordinates with imaginary quantum dimensions. We will investigate a unified quantum field theory of this type. We will create a unified theory of the Standard Model and Quantum Gravity. The unified theory has the following features:

1. Consistency with the current Standard Model in all respects at current energies up to current maximal energies of several TeV.

2. Consistency with the classical Theory of Gravity. Classical gravity is the "large distance", "low energy" limit of the theory.

3. General relativistic covariance of the field equations, Lorentz invariance of the S matrix, and unitarity as required in physically acceptable theories.

4. The unified theory is divergence-free – it contains no infinities.

5. All interactions are modified by a short distance, high-energy, substructure that begins at some energy presumably much above currently accessible energies. The energy scale could be set by the Planck scale ($10^{19}$ GeV) or could be a much lower energy such as $10^5$ GeV.

6. It allows modifications of the Standard Model such as further unification through broken higher symmetries without losing features 1 – 5 above.

7. It predicts a number of new phenomena at extremely short distances such as ultra-relativistic dilepton resonances consisting of two leptons of the same charge. An example of this type of bound state is a bound resonance consisting of two electrons. These dilepton states could be created at ultrahigh energies by penetrating the repulsive Coulomb barrier. Inside the barrier near $r = 0$ the modified Coulomb potential is a linear potential. A dilepton bound state is highly unstable with a large

decay rate due to quantum tunneling through the Coulomb barrier. Other exotic resonances are also possible.

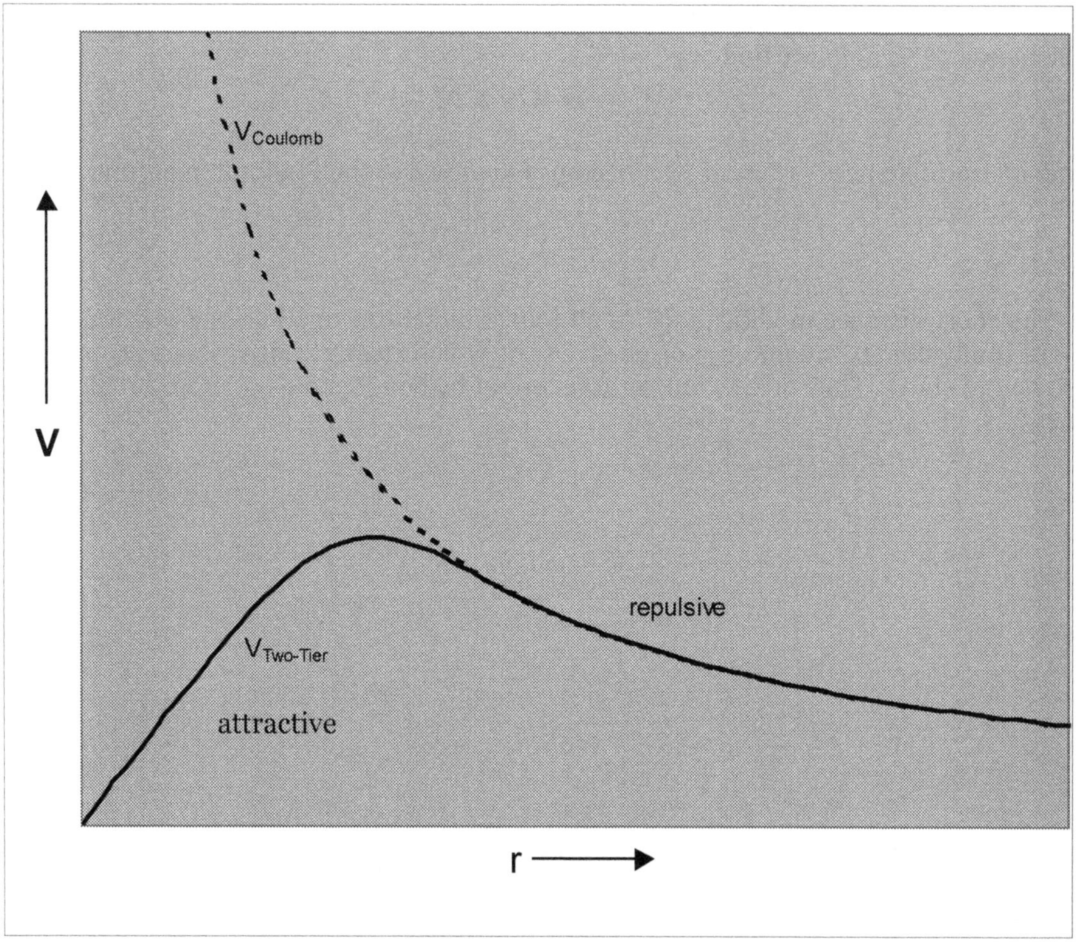


Figure 1.1. Modified electromagnetic Coulomb force between two particles with charges of the same sign (eg. two electrons) in two-tier QED and the two-tier Standard Model. The potential is repulsive at long distances and attractive at short distances. The potential becomes linear near the origin r = 0 opening up the possibility of unstable doubly charged resonances.

The curve in Fig. 1.1 is the two-tier potential between two particles of the same sign. The two-tier electromagnetic potential between two singly charged particles of the same sign is:

$$V_{new} = \alpha\Phi(M_c^2\pi r^2)/r \tag{1.18}$$

where $\alpha$ is the fine structure constant, $\Phi(z)$ is the error function, $M_c$ is the mass setting the scale of the new short distance behavior and r is the radial distance. At small distances ($\pi r^2 \ll M_c^{-2}$) we find

$$V_{new} \rightarrow \alpha 2\sqrt{\pi}\, M_c^2 r \tag{1.19}$$

and at large distances ($\pi r^2 \gg M_c^{-2}$) it becomes identical to the Coulomb potential:

$$V_{new} \rightarrow V_{Coul} = \alpha/r \tag{1.20}$$

8. The short distance modifications of all four interactions open up the possibility of a more fundamental, smaller set of particles, of which the currently observed particles are bound states. Thus quarks and leptons might be bound states of more fundamental particles.

# 2. The Standard Model, Gravity and Superstring Theories

In this chapter we consider aspects of the Standard Model and Gravitation, and Superstring theories, that are relevant for the discussion of our new unified theory.

## Standard Model

The Standard Model unites the Electroweak interactions and the Strong interactions in a successful theory that appears to be consistent with the known experimental data at all accessible energies. The theory satisfies the unitarity condition with positive probabilities adding to unity, Lorentz invariance, and renormalizability. There are several variants and proposed extensions of the Standard Model. But the major aspects of the theory are common to all variants.

The most important technical issue that faced the development of the Standard Model was the question of the renormalizability of the unification of the electromagnetic interactions and the weak interactions in the Electroweak Theory sector. The issue was resolved by the proof of renormalizability by 't Hooft in 1971.

At first glance Electroweak Theory, and its electromagnetic sector, Quantum Electrodynamics (QED), appear to have divergences when perturbative calculations are made of transition probabilities, or of physical quantities such as the anomalous magnetic moment of the electron or muon. However these divergences can be isolated into a finite number of infinite quantities (renormalization theory) which in turn can be absorbed into redefinitions of fundamental parameters of the theory (such as the "bare" electric charge or the "bare" mass of the electron.)

For example in QED the the "bare" charge $e_0$ is renormalized by infinite factors to give the "physical" charge e: $e = Z_1^{-1}Z_2\sqrt{Z_3}e_0$ where $Z_1$, $Z_2$, and $Z_3$ are divergent renormalization constants. Numerous authors over the last fifty years or so have commented on the "unnaturalness" of renormalization which, after all, amounts to multiplying infinity by zero ($e_0$) to obtain a finite observable number, the electric charge e in this case.

The quantum field theory formalism that we shall develop resolves these issues by being divergence-free – no infinities appear because of a short distance modification that appears in all particle propagators. The result is a logically satisfactory theory that avoids divergences independent of the details of the interactions and symmetries, and, in fact, allows a wider range of interactions such as interactions with derivative couplings, which were previously totally unrenormalizable.

## Quantization of Gravity

Classical gravity began with Newton's theory of gravity, which still remains an acceptable theory for most phenomena involving small masses and velocities. In the early twentieth century A. Einstein developed a new theory of gravitation based on geometrical concepts. This theory, the General Theory of Relativity, has been tested and found to be correct as far as we can determine at the classical level.

There have been many attempts to develop a quantum version of the General Theory of Relativity. While part of the overall framework of such a quantum theory is known, all attempts to develop it within the framework of conventional quantum field theory have failed due to the infinite number of divergences that appear when calculations are attempted in perturbation theory.[5] A theory with an infinite number of divergences cannot be handled with the renormalization procedures used to tame the divergences of QED, Yang-Mills theories, or the Electroweak Theory; and thus does not have predictive power. The gravitational sector of the unified theory that we will develop does not have these divergences and is in fact divergence-free.

## Correspondence Principle

The development of Quantum Mechanics was guided by a principle developed by Bohr called the *Correspondence Principle.* Simply put, this principle states that quantum mechanical systems must approach the behavior of classical systems as the features of the quantum mechanical system become much larger than the quantum of action – Planck's constant h. Classical mechanics is thus the large scale limit of quantum mechanics.

Similarly, the Theory of Special Relativity also has a sort of correspondence principle – the predictions of the theory of Special Relativity for a system must approach the predictions of classical mechanics in the limit that all the velocities of the system become much smaller than the speed of light. Thus classical mechanics is a limiting case of Special Relativity.

Just as these twentieth century theories have a type of correspondence principle in which they must approximate an earlier theory in a certain limit, any new theory of elementary particles must approximate the Standard Model in the currently known range of energies. After all the Standard Model works! Therefore it must have an element of physical truth and a more fundamental theory must embody that truth in the explored range of experimental energies.

---

[5] See for example B. S. DeWitt, Phys. Rev. **162**, 1239 (1967), and R. P. Feynman, Acta Physica Polonica **24**, 697 (1963).

Furthermore the extreme accuracy of QED calculations[6] must be matched by any theory purporting to account for elementary particle interactions. At the moment, and apparently for the foreseeable future, Superstring theories have difficulties when attempts are made to find the Standard Model, or something like it, as an approximation at current energies. Capturing the accuracy of QED predictions would seem to be in the distant future, if at all, in Superstring theories.

## String and Superstring Theories

W. Pauli once remarked, "Man should not join together that which God has put asunder." In apparent contradiction (perhaps) to this dictum Superstring theory attempts to unite fermions (half-integer spin particles) and bosons (integer-spin particles) within a larger symmetry so that they can, in a sense, be rotated into one another. While support for this symmetry is currently lacking in elementary particle physics, there may be evidence for supersymmetry in nuclear physics.

However the price for supersymmetry in particle physics appears to be quite high: a large number of dimensions and a large number of particles. No evidence exists at the time of this writing for more than four space-time dimensions or for the supersymmetric partners of the known elementary particles. Thus, at best, supersymmetry is for the future when experimental energies reach energies where supersymmetric features are unequivocally seen (if in fact they exist.)

The theory of strings – two-dimensional substructures that constitute elementary particles – is grounded in phenomenology – more particularly, in the Veneziano-Suzuki formula for scattering amplitudes. Y. Nambu and T. Goto developed a vibrating string theory that accounted for the form of the Veneziano-Suzuki formula. So there appears to be some justification for a string model of elementary particles. In this model the elementary particles, which appear as structure-less and point-like at lower energies, can be seen to have a string-like structure at very high energies.

Thus a sub-structure for elementary particles may have some experimental justification.

## Logic Does Not Necessarily Bring Progress in Physics

An often-expressed hope in the last two decades of the twentieth century was that some unique Superstring theory would emerge, and that this theory would prove to be the only logically reasonable theory. After thirty years of effort by a large group of extremely talented physicists (perhaps larger than any group of physics theorists devoted to one topic since the Manhattan project) this hope is yet to be realized.

Historically, the movement from level to deeper level to yet deeper level in Physics has not been the result of logical inquiry but rather have been the result of new experiments confronting existing theory wherein the existing theory is found to be

---

[6] T. Kinoshita, "The Fine Structure Constant", Cornell Univ. Preprint CLNS 96/1406 (1996).

wanting.[7] A rational physicist in the eighteenth century would not have conjectured the theory of special relativity as a logical possibility: "Why should mechanics change when the speed of an object becomes large?" A rational physicist in the nineteenth century would not have conjectured the theory of quantum mechanics as a logical possibility: "Why should it be impossible to measure the momentum and position of a particle with arbitrary precision?" These questions did not, and could not have arisen, in a rational, "down to earth" scientist. Furthermore, despite knowing that these theories are correct we do not know WHY they are implemented in Nature. We only know that they are.[8]

With this historical perspective in mind it appears that we should not expect to arrive at the form of the next deeper level of Physics through reason alone. Nature will most likely surprise us again (and again). Thus experiment is our teacher as we learn more of the nature of matter and space-time.

## Goal: Unified Theory Without Renormalization Issues

There are a number of issues confronting elementary particle physics today. Issues such as CP violation, the nature of dark matter and of dark energy, the form of the symmetries of Nature, and the origin of the "numerous" particles and constants in the Standard Model. Practically speaking, perhaps the most important problem at this time is the development of a renormalizable unified theory of all the interactions. We have seen that we can cope with the incomplete unification in the Standard Model between the electroweak interactions and the strong interaction. The Standard Model is renormalizable and thus we can make perturbation theory calculations with confidence and compare the results with experiment. But we cannot make calculations in quantum gravity with confidence. As a result Planck scale physics is totally speculative and we cannot understand the nature of the Big Bang when the universe was contained within a region the size of the Planck length.

With these issues in mind we have developed a unified theory of all the known interactions that is divergence-free. Thus we can perform calculations to any order of perturbation theory in any sector and obtain finite results that can be compared with experiment. The theory has the Standard Model and classical General Relativity as "low energy" limits. At high energies a string-like sub-structure generates a smooth high-energy limit that eliminates the divergences in perturbation theory. Thus there is a correspondence principle for our theory – it has the right "low energy" limits.

The nature of the string-like sub-structure is not dependent on the details of the interactions and symmetries. The interactions in the Standard Model and quantum gravity are can have any polynomial form (in the quantum fields). A variety of other interactions (such as those with derivative couplings) are allowed in this approach

---

[7] This historical process of Physics is described in some detail in the book Blaha(2002) by this author.

[8] A possible answer as to why Nature is quantum in nature has recently been proposed by Blaha (2005) based on Gödel's Theorem.

which do not affect the finiteness of the theory. Thus the range of possible extensions of the Standard Model is significantly widened.

# 3. Quantization of Coordinate Systems

## Non-commuting Coordinates

Field theories with non-commuting coordinates are currently an active field of study.[9] Investigators are studying gauge theories, and in particular Quantum Electrodynamics, with non-commuting coordinates. Non-commuting coordinates are usually implemented quantum mechanically by positing non-zero commutators for coordinates:

$$[x^i, x^j] = i\theta^{ij} \tag{3.1}$$

*New Approach to Non-Commuting Coordinates*

*In this book we will consider an alternative approach that postulates a q-number coordinate system $X^\mu$ with which all particle fields are defined. This coordinate system is realized as a mapping from a more fundamental c-number coordinate system $y^\nu$, which we will call the subspace for want of a better term. We will treat $X^\mu$ as a vector of quantum fields, thus realizing a new type of non-commutative coordinates at unequal subspace times.*

This approach is radically different from the non-commutative coordinate realizations hitherto discussed in the literature. It has a number of beneficial results to recommend it – the main result is the finiteness of quantum field theories that are defined within its framework. We will explore some of these results in the following chapters.

The $X^\mu$ coordinate system, as we define it, has a c-number real part and a q-number imaginary part. Thus particle fields which are normally defined on four-dimensional real space-time will now be defined on a complex four-dimensional space-time where four imaginary dimensions will appear as *Quantum Dimensions*™ embodied in a vector quantum field $Y^\mu(y)$.

---

[9] M. R. Douglas and N. A. Nekrasov, Rev. Mod. Phys. **73**, 977 (2002) and references therein; J. Harvey, hep-th/0102076; M. Hamanaka and K. Toda, hep-th/0211148; N. Seiberg and E. Witten, hep-th/9908142; R. J. Szabo, hep-th/0109162; G. Berrino, S. L. Cacciatori, A. Celi, L. Martucci, and A. Vicini, hep-th/0210171; S. Godfrey and A. Doncheski, DESY eprint 02-195; M. Caravati, A. Devoto, and W. W. Repko, hep-th/0211463; and references within these papers.

$$X_\mu(y) = y_\mu + i\, Y_\mu(y)/M_c^2$$

The $Y^\mu(y)$ field is a function of the subspace y coordinates. The real part of the space-time dimensions will be taken to be the subspace y coordinates.[10]

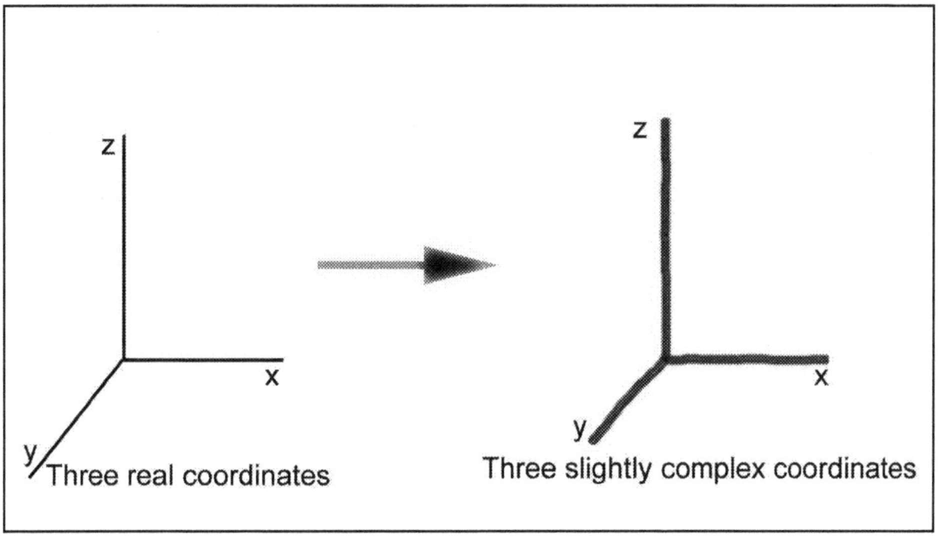


Figure 3.1. The change from purely real space to a slightly complex space with imaginary quantum fluctuations for each spatial axis in the Coulomb gauge of the Y field.

The imaginary part of space-time (which has not been experimentally seen) will simply be the quantum fluctuations of a massless vector quantum field that are suppressed further by a large mass scale – perhaps of the order of the Planck mass – that reduces the imaginary Quantum Dimensions™ to the infinitesimal. The effects of Quantum Dimensions™ only become appreciable in quantum field theory at energies of the order of $M_c$. At these energies the exponential Gaussian factor in each particle (and ghost) propagator that is generated by the Quantum Dimensions™ serves to make perturbation theory calculations ultra-violet finite – including calculations in Quantum Gravity.

The formalism that we will describe introduces a new form of interaction that does not have the form of the simple polynomial interactions that have hitherto

[10] In a deeper theory the real part might also be a quantum field that undergoes a condensation to generate c-number coordinates. We will not consider this possibility in this book.

dominated quantum field theories. This form of interaction takes place via the composition of quantum fields and can be called a *Dimensional Interaction*™ or an *Interdimensional Interaction*™ since it affects particle behavior through Quantum Dimensions™.[11]

## Quantization Using a C-Number $X^\mu$

We will begin by considering the case of a scalar quantum field theory. We assume a real underlying y subspace. Since $X^\mu$ is a set of coordinates, we choose to define a scalar field $\phi$ as a function of $X^\mu$, which in turn is a function of the $y^\nu$ coordinates. We will provisionally second quantize $\phi$ treating $X^\mu$ as c-number coordinates using a conventional approach.[12]

We assume a Lagrangian specified by eq. 1.10 that leads to the Klein-Gordon equation eq. 1.15. For that Lagrangian formulation, the momentum conjugate to $\phi$ is:

$$\pi_\phi = \partial L_F \ / \ \partial\phi' \ \equiv \ \partial L_F \ / \ \partial(\partial\phi/\partial X^0) \tag{3.2}$$

Following the canonical quantization procedure, $\pi$ and $\phi$ become hermitian operators with equal time ($X^0 = X^{0\prime}$) commutation rules:

$$[\phi(X), \phi(X')] = [\pi_\phi(X), \pi_\phi(X')] = 0 \tag{3.3}$$

$$[\pi_\phi(X), \phi(X')] = -i\,\delta^3(\mathbf{X} - \mathbf{X}') \tag{3.4}$$

The hamiltonian is defined by eq. A.112. (Appendix A contains a detailed development of the formalism for the scalar particle case. It was placed there because there are many formal similarities to conventional quantum field and this approach allows us to proceed more quickly to the main points of difference between conventional quantum field theory and two-tier quantum field theory in the present chapter. Appendix A also describes a new type of method – the composition of extrema – for the Calculus of Variations. *Equations numbered A.xxx are in Appendix A.*) We assume a metric $\eta_{\mu\nu}$ where $\eta_{00} = +1$, $\eta_{0i} = 0$, and $\eta_{ij} = -1$ for i, j = 1,2,3.

---

[11] See the back of title page for the rationale for, and the details of the authorization to use, these trademarked terms.

[12] Some texts are: Bogoliubov, N. N., Shirkov, D. V., *Introduction to the Theory of Quantized Fields* (Wiley-Interscience Publishers Inc., New York, 1959); Bjorken, J. D., Drell, S. D., *Relativistic Quantum Fields* (McGraw-Hill, New York, 1965); Huang, K., *Quarks, Leptons & Gauge Fields Second Edition* (World Scientific, River Edge, NJ, 1992); Kaku, M., *Quantum Field Theory* (Oxford University Press, New York, 1993); Weinberg, S., *The Quantum Theory of Fields* (Cambridge University Press, New York, 1995).

The standard Fourier expansion of the solution to the Klein-Gordon equation (eq. A.34) is:

$$\phi(X) = \int d^3p \, N_m(p) \, [a(p) \, e^{-ip\cdot X} + a^\dagger(p) \, e^{ip\cdot X}] \tag{3.5}$$

where

$$N_m(p) = [(2\pi)^3 2\omega_p]^{-\frac{1}{2}} \tag{3.6}$$

and

$$\omega_p = (\mathbf{p}^2 + m^2)^{\frac{1}{2}} \tag{3.7}$$

The commutation relations of the Fourier coefficient operators are:

$$[a(p), a^\dagger(p')] = \delta^3(\mathbf{p} - \mathbf{p}') \tag{3.8}$$

$$[a^\dagger(p), a^\dagger(p')] = [a(p), a(p')] = 0 \tag{3.9}$$

The reader will recognize the quantization procedure is formally identical to the standard canonical quantization procedure of a free scalar quantum field.

In the case of spin ½, spin 1 and spin 2 fields the standard quantization procedure *in terms of the X coordinate system* can also be followed in a way similar to the procedure in standard texts. We will see these quantization procedures in the following chapters. In the next section we will quantize the transformation from the y coordinate system to the X coordinate system.

*The procedures developed in this section and the following sections may disturb some readers since we are placing operators with Dirac delta functions and using other unusual operator expressions. These concerns should be put at rest when we show that a path integral formulation presented later gives precisely the same results as the present development.*

## Coordinate Quantization

In this section we will quantize the coordinates $X^\mu$ as a vector field defined on a fundamental c-number coordinate system $y^\nu$ of the same dimensionality. We will assume the $y^\nu$ space is a "normal" flat Minkowski space with three spatial and one time dimensions. Generalizations to spaces with more dimensions are straightforward but will not be considered here.

Thus we will assume $X^\mu$ has three spatial dimensions and one time dimension. For reasons primarily of simplicity (primarily to avoid multiple time coordinates) we will assume the $X^\mu$ fields are similar to the free electromagnetic vector potential $A^\mu$ with the Lagrangian:

$$\mathscr{L}_C = +\frac{1}{4} M_c^4 F^{\mu\nu} F_{\mu\nu} \qquad (3.10)$$

$$F_{\mu\nu} = \partial X_\mu / \partial y^\nu - \partial X_\nu / \partial y^\mu \qquad (3.11)$$

where $M_c^4$ is a mass scale to the fourth power that is required on dimensional grounds and serves to set the scale for new Physics as we will see later. *Note the sign in eq. 3.10 is not negative – superficially contrary to the conventional electromagnetic Lagrangian. The reason for this difference is that the field part of $X^\mu$ is imaginary.* Thus $\mathscr{L}_C$ winds up having the correct sign after taking account of the factors of i in the field strength $F_{\mu\nu}$.

We assume $X^\mu$ is complex[13] with the form:

$$X_\mu(y) = y_\mu + i\, Y_\mu(y)/M_c^2 \qquad (3.12)$$

where $Y_\mu(y)$ is a quantum field, $M_c$ is a mass scale, and the real part is the c-number 4-vector $y_\mu$. If $X^\mu$ has this form, then

$$F_{\mu\nu} = i\,(\partial Y_\mu / \partial y^\nu - \partial Y_\nu / \partial y^\mu)/M_c^2 \qquad (3.13)$$

Defining

$$F_{Y\mu\nu} = (\partial Y_\mu / \partial y^\nu - \partial Y_\nu / \partial y^\mu) \qquad (3.14)$$

we see the Lagrangian assumes the form of the conventional electromagnetic Lagrangian:

$$\mathscr{L}_C = -\frac{1}{4} F_Y^{\mu\nu} F_{Y\mu\nu} \qquad (3.15)$$

This Lagrangian can be used to develop field equations and a canonical quantization that is completely analogous to Quantum Electrodynamics.

[13] Theories of quantum mechanics, and quantum fields, in complex and quaternion spaces have been considered by numerous authors. For example see C. M. Bender, D. C. Brody and H. F. Jones, "Complex Extension of Quantum Mechanics" Phys. Rev. Letters **89**, 270401-1 (2002) and references therein; S. L. Adler and A. C. Millard, "Generalized Quantum Dynamics as Pre-Quantum Mechanics", Princeton Univ. preprint arXiv:hep-th/9508076 (1995) and references therein. These theories are all very different from the theories presented herein.

*Gauge Invariance*

The gauge invariance of the Lagrangian allows us to choose a convenient gauge. The gauge invariance of the full Lagrangian

$$\mathscr{L}_s = \mathscr{L}_F(\phi(X), \partial\phi/\partial X^\mu)\, J + \mathscr{L}_C(X^\mu(y), \partial X^\mu(y)/\partial y^\nu) \qquad (A.96)$$

is based on the standard gauge invariance of $\mathscr{L}_C$, and the gauge invariance of $J\mathscr{L}_F$ in the form of translational invariance

$$X^\mu(y) \rightarrow X^\mu(y) + \delta X^\mu(y) \qquad (A.97)$$

for the special case of a translation of X with the form of a gauge transformation:

$$\delta X^\mu(y) = \partial\Lambda(y)/\partial y_\mu$$

In this case eq. A.106 implies

$$\int d^4y\, \Lambda(y)\, \partial\, [\, J\, \partial/\partial X^\mu\, \mathscr{T}_{F\mu\nu}\, ]/\partial y_\nu = 0$$

after a partial integration and so gives the differential conservation law:

$$\partial\, [\, J\, \partial/\partial X^\mu\, \mathscr{T}_{F\mu\nu}\, ]/\partial y_\nu = 0 \qquad (3.16)$$

since $\Lambda(y)$ is arbitrary. This conservation law is trivially obeyed since, by eq. A.108:

$$\partial/\partial X^\mu\, \mathscr{T}_{F\mu\nu} = 0 \qquad (A.108)$$

Thus translational invariance in the $\mathscr{L}_F$ sector together with standard gauge invariance in the $\mathscr{L}_C$ sector automatically guarantees Y field gauge invariance of the total Lagrangian. Basically we use the separate invariance of each term of

$$L = \int d^4y\, [\mathscr{L}_F J + \mathscr{L}_C] = \int d^4X\, \mathscr{L}_F + \int d^4y\, \mathscr{L}_C = L_F + L_C$$

under a constant translation $X^\mu \rightarrow X^\mu + \delta X^\mu$ where $\delta X^\mu$ is constant to establish eq. A.108. Then we consider a position dependent translation/gauge transformation to derive eq. 3.16, which taken together with eq. A.108, establishes the invariance under the position dependent translation/gauge transformation eq. A.97.

An alternate approach that leads to the same result is to start with the particle part of the Lagrangian $L_F$ rewritten to be invariant under general coordinate transformations as it must when we generalize to include General Relativity. Since position dependent translations are a form of general coordinate transformation the full theory must be invariant under position dependent translations due to invariance under general coordinate transformations.

Having established invariance under gauge transformations we now choose to use the most convenient gauge – the Coulomb gauge[14]:

$$\partial Y^i/\partial y^i = 0 \tag{3.17a}$$

which, in the absence of external sources, allows us to set

$$Y^0 = 0 \tag{3.17b}$$

since $Y^0$ does not have a canonically conjugate momentum. A conventional treatment leads to the equal time commutation relations:

$$[Y^\mu(\mathbf{y}, y^0), Y^\nu(\mathbf{y}', y^0)] = [\pi^\mu(\mathbf{y}, y^0), \pi^\nu(\mathbf{y}', y^0)] = 0 \tag{3.18}$$

$$[\pi^j(\mathbf{y}, y^0), Y_k(\mathbf{y}', y^0)] = -i\,\delta^{tr}{}_{jk}(\mathbf{y} - \mathbf{y}') \tag{3.19}$$

(Note the locations of the j indexes in eq. 3.19 introduce a minus sign.) where

$$\pi^k = \partial \mathscr{L}_C \;/\; \partial Y_k' \tag{3.20}$$

$$\pi^0 = 0 \tag{3.21}$$

$$\delta^{tr}{}_{jk}(\mathbf{y} - \mathbf{y}') = \int d^3k\, e^{i\,\mathbf{k}\cdot(\mathbf{y}-\mathbf{y}')}(\delta_{jk} - k_j k_k/\mathbf{k}^2)/(2\pi)^3 \tag{3.22}$$

$$Y_k' = \partial Y_k/\partial y^0 \tag{3.23}$$

---

[14] It is also possible to quantize using an indefinite metric that preserves manifest Lorentz covariance as was done by Gupta and Bleuler for the electromagnetic field. We will use the Gupta-Bleuler approach later to establish covariance under special relativity later. Now we opt for manifest positivity and use the Coulomb gauge.

The Coulomb gauge reveals the two degrees of freedom that are present in the vector potential. The Fourier expansion of the vector potential is:

$$Y^i(y) = \int d^3k\, N_0(k) \sum_{\lambda=1}^{2} \varepsilon^i(k, \lambda)[a(k,\lambda)\, e^{-ik\cdot y} + a^\dagger(k,\lambda)\, e^{ik\cdot y}] \qquad (3.24)$$

where

$$N_0(k) = [(2\pi)^3 2\omega_k]^{-1/2} \qquad (3.25)$$

and (since m = 0)

$$\omega_k = (\mathbf{k}^2)^{1/2} = k^0 \qquad (3.26)$$

with $\vec{\epsilon}(k, \lambda)$ being the polarization unit vectors for $\lambda = 1,2$ and $k^\mu k_\mu = 0$.

The commutation relations of the Fourier coefficient operators are:

$$[a(k,\lambda),\, a^\dagger(k',\lambda')] = \delta_{\lambda\lambda'}\delta^3(\mathbf{k} - \mathbf{k}') \qquad (3.27)$$

$$[a^\dagger(k,\lambda),\, a^\dagger(k',\lambda')] = [a(k,\lambda),\, a(k',\lambda')] = 0 \qquad (3.28)$$

and the polarization vectors satisfy

$$\sum_{\lambda=1}^{2} \varepsilon_i(k, \lambda)\varepsilon_j(k, \lambda) = (\delta_{ij} - k_i k_j/\mathbf{k}^2) \qquad (3.29)$$

It will be convenient to divide the Y field into positive and negative frequency parts:

$$Y^+{}_i(y) = \int d^3k\, N_0(k) \sum_{\lambda=1}^{2} \varepsilon_i(k, \lambda)\, a(k,\lambda)\, e^{-ik\cdot y} \qquad (3.30)$$

and

$$Y^-{}_i(y) = \int d^3k\, N_0(k) \sum_{\lambda=1}^{2} \varepsilon_i(k, \lambda)\, a^\dagger(k,\lambda)\, e^{ik\cdot y} \qquad (3.31)$$

For later use we note the commutator between the positive and negative frequency parts is:

$$[\, Y^-{}_j(y_1),\, Y^+{}_k(y_2)] = -\int d^3k\, e^{ik\cdot(y_1 - y_2)}\, (\delta_{jk} - k_j k_k/\mathbf{k}^2)/[(2\pi)^3 2\omega_k] \qquad (3.32)$$

## Bare $\phi$ Particle States

We now turn to the $\phi$ particle states. The creation and annihilation operators can be used to define "bare" free particle states. Bare free particle states are states that are not dressed with coherent states of Y quanta. For example a bare one-particle state of momentum p is

$$|p> = a^\dagger(p)\,|0_\phi> \tag{3.33}$$

with corresponding bare bra state

$$<p| = <0_\phi|\,a(p) \tag{3.34}$$

where the vacuum is defined as usual:

$$a(p)\,|0_\phi> = 0 \tag{3.35}$$

$$<0_\phi|\,a^\dagger(p) = 0 \tag{3.36}$$

Multi-particle bare states can also be defined in the conventional way with products of creation and annihilation operators applied to the vacuum.

## Y Fock Space Imaginary Coordinate States

States can also be defines for the quantized Y field. These states will be similar in form to electromagnetic photon states but play a different role in our approach since they are in fact coordinate excitation states for the imaginary part of $X^\mu$. Thus the scalar field (and other particle fields) will exist in a real four-dimensional space with quantum excitations into imaginary Quantum Dimensions™. These excitations become significant at high energies. At the low energies with which we are familiar, space-time appears real; at very high energies space-time becomes slightly complex.

There are two types of imaginary coordinate excitations: 1.) Quantum excitations into Fock states consisting of superpositions of states with a definite finite number of Y "particles" and 2.) Imaginary coordinate excitations into coherent Y states with an "infinite" number of particles. Coherent states can be viewed as representing "classical" fields.

In this section we will consider Y field states with a definite number of excitations ("particles"). The creation and annihilation operators of the Y field can be used to define free particle states. For example a one particle state can be defined by

$$|k, \lambda> = a^\dagger(k, \lambda)\,|0_Y> \tag{3.37}$$

with corresponding bra state

$$<k, \lambda| = <0_Y| a(k, \lambda) \qquad (3.38)$$

where the "coordinate vacuum" is defined as usual:

$$a(k, \lambda) | 0_Y> = 0 \qquad (3.39)$$

$$<0_Y | a^\dagger(k, \lambda) = 0 \qquad (3.40)$$

Multi-particle states can also be defined in the conventional way with products of the creation and annihilation operators applied to the vacuum. The set of all states containing a finite number of "particles" constitutes a Fock space.

*A state with a finite number of Y "particles" represents a quantum fluctuation into imaginary Quantum Dimensions™. Such states do not appear in two-tier quantum field theory since the Y field is a free field and has no source. Thus they appear only as part of normal particles. A normal particle, such as a $\phi$ particle, has a coherent state of Y quanta associated with it, which play a role in interactions. The Y coherent state part of a normal particle can be viewed as boring an infinitesimal "hole" into an extra pair of imaginary dimensions in a neighborhood of the particle of a radial extent set by the length* $M_c^{-1}$.

## Y Coherent Imaginary Coordinate States

Coherent Y states bring us closer what we might consider to be "classical" imaginary dimensions – dimensions that we can, in principle, experience as we do normal dimensions. Let us define the coherent state[15]

$$| y, p> = e^{-\mathbf{p}\cdot\mathbf{Y}^-(y)/M_c^2} | 0_Y> \qquad (3.41)$$

This state is an eigenstate of the coordinate operator $Y^+(y')$:

$$Y^+_j(y_1) \ |y_2, p> = -[Y^+_j(y_1), \mathbf{p}\cdot\mathbf{Y}^-(y_2)]/M_c^2 \ |y, p> \qquad (3.42)$$

$$= -\int d^3k\, [N_0(k)]^2\, e^{ik\cdot(y_2 - y_1)}\, (p_j - k_j\mathbf{p}\cdot\mathbf{k}/\mathbf{k}^2)/M_c^2 \ |y, p> \qquad (3.43)$$

[15] Coherent states are well known in the physics literature. See for example T. W. B. Kibble, J. Math. Phys. **9**, 315 (1968) and references therein; V. Chung, Phys. Rev. **140**, B1110 (1965); J. R. Klauder, J. McKenna, and E. J. Woods, J. Math. Phys. **7**, 822 (1966) and references therein.

$$= p^i\Delta_{Tij}(y_1 - y_2)/M_c^2 \; |y, p> \tag{3.44}$$

where $p^i\Delta_{Tij}(y_1 - y_2)/M_c^2$ is the eigenvalue of $Y^+{}_j(y_1)$. As we will see in the next chapter, the eigenvalue of $Y^+$ becomes large as $(y_1 - y_2)^2 \to 0$. Thus the imaginary Quantum Dimensions™ become significant at very short distances, and significantly modify the high-energy behavior of quantum field theories. In particular, Quantum Dimensions™ have a significant effect when

$$(y_1 - y_2)^2 \lessapprox (4\pi^2 M_c^2)^{-1} \tag{3.45}$$

according to eq. 4.13 in the next chapter. We are assuming the mass scale $M_c$ is very large – perhaps of the order of the Planck mass ($1.221 \times 10^{19}$ GeV/$c^2$). Thus imaginary Quantum Dimensions™ are far from detectable in today's "low" energy experiments. Their effect are significant in the analysis of the first instants after the Big Bang.[16]

*The Dynamical Generation of New Dimensions*

Effectively, the imaginary dimensions that we have constructed raise the total number of real and Quantum Dimensions™ to 8 with 6 space dimensions and two time dimensions. As we will see later the requirement of gauge invariance for the quantized Y field reduces the number of time dimensions to one and constrains the six space dimensions to five degrees of freedom giving a 5+1 dimensional space. Since X is a function of y we can also view the four dimensional world that we live in as a four-dimensional surface in a 6-dimensional space-time.

*Generation of Quantum Dimensions™ by the $\phi$(X) field*

The $\phi(X)$ field generates Quantum Dimensions™ via coherent states from the vacuum. From eq. 3.5 and 3.12 we see

$$\phi(X) = \int d^3p \, N_m(p) \, [a(p) \, e^{-ip\cdot(y + iY/M_c^2)} + a^\dagger(p) \, e^{ip\cdot(y + iY/M_c^2)}] \tag{3.46}$$

with the result

$$\phi(X)|0> = \int d^3p \, N_m(p) \, a^\dagger(p) \, e^{ip\cdot(y + iY/M_c^2)}|0> \tag{3.47}$$

is a superposition of coherent Y states plus one scalar particle. The vacuum state $|0>$ is the product of the $\phi$ and Y vacuum states $|0> = |0_Y>|0_\phi>$. We will use $|0>$ in most of the following discussions.

---

[16] Blaha (2004).

We can also define coherent Y states with total momentum q using the expression:

$$|q\,Y\rangle = \int d^4y\, e^{iq\cdot X(y)}|0\rangle = \int d^4y\, e^{iq\cdot(y + iY/M_c^2)}|0\rangle \tag{3.48}$$

Expanding the Y part of the exponential in eq. 3.48 gives

$$|q\,Y\rangle = \sum_{n=0}^{\infty}(-1)^n(n!)^{-1}\prod_{j=1}^{n}\left(\int d^3k_j N_0(k_j)\right)\delta^4\!\left(q - \sum_{s=1}^{n} k_s\right)\prod_{r=1}^{n}\sum_{\lambda_r=1}^{2} \mathbf{q}\cdot \varepsilon(k_r, \lambda_r)\, a^\dagger(k_r,\lambda_r)|0\rangle \tag{3.49}$$

which indicates that the sum of the Y particle momenta for each term in the expansion is q.

## Hamiltonian for Particle and Coordinate States

The hamiltonian for the separable (field hamiltonian term separate from the Y hamiltonian term – see Appendix A), coordinate quantized, scalar quantum field theory is:

$$H_s = \int d^3y\, \mathscr{H}_s \tag{A.79}$$

with

$$\mathscr{H}_s = J\mathscr{H}_F + \mathscr{H}_C \tag{A.82}$$

$$\mathscr{H}_F(\phi(X), \pi_\phi, \partial\phi/\partial X^i) = \pi_\phi\, \phi' - \mathscr{L}_F \tag{A.83}$$

$$\mathscr{H}_C(X^\mu(y), \pi_X^{\ \mu}, \partial X^\mu(y)/\partial y^j, y^\nu) = \pi_X^{\ \mu}\, X_\mu' - \mathscr{L}_C \tag{A.84}$$

$$\mathscr{L}_F = ½\,[\,(\partial\phi/\partial X^\nu)^2 - m^2\phi^2\,] \tag{A.33}$$

$$\mathscr{L}_C = -¼\, M_c^4 F_Y^{\ \mu\nu} F_{Y\mu\nu} \tag{3.15}$$

We note

$$\mathscr{H}_F = ½\,[\pi_\phi^{\,2} + (\partial\phi/\partial X^i)^2 + m^2\phi^2] \tag{3.50}$$

is the conventional scalar particle hamiltonian when viewed as a function of the X coordinates. $\mathscr{H}_C$ has the same form as the conventional electromagnetic hamiltonian when eq. 3.12 is used to specify X in terms of the Y fields.

$$\mathcal{H}_C = \tfrac{1}{2}\,(E_Y^{\ 2} + B_Y^{\ 2}) \tag{3.51}$$

where

$$E_Y^{\ i} = -\partial Y^i/\partial y^0 \tag{3.52}$$

$$B_Y^{\ i} = \varepsilon^{ijk}\,\partial Y_j/\partial y^k \tag{3.53}$$

Using the fourier expansions of $\phi$ and $X^\mu$ (eqs. 3.5 and 3.24) we obtain the following expression for the normal-ordered hamiltonian $H_S$:

$$P_s^{\ 0} \equiv H_s = \int :\mathcal{H}_s:\, d^3y \tag{3.54}$$

$$H_s = \int d^3p\,(\mathbf{p}^2 + m^2)^{1/2} a^\dagger(p)a(p) + \int d^3k \sum_{\lambda=1}^{2} (\mathbf{k}^2)^{1/2}\, a^\dagger(k, \lambda)a(k, \lambda) \tag{3.55}$$

where : : indicates normal ordering and where we perform a functional integration over X (Note the Jacobian is present within $\mathcal{H}_S$.) for the particle part of the hamiltonian $\mathcal{H}_F$. The hamiltonian is manifestly positive definite.

The spatial momentum is specified by

$$P_s^{\ j} = -\int d^3X\, :\pi_\phi(X)\partial\phi(X)/\partial X_j: + \int d^3y\, :E_Y^{\ i}\partial Y^i/\partial y_j: \tag{3.56}$$

$$= \int d^3p\, p^j\, a^\dagger(p)a(p) + \int d^3k \sum_{\lambda=1}^{2} k^j\, a^\dagger(k, \lambda)a(k, \lambda) \tag{3.57}$$

where the first term in eq. 3.57 follows because of $\int d^3X$ in eq. 3.56. The momentum operator generates displacements in $\phi$

$$[P_s^{\ \mu}, \phi(X)] = -\,i\partial\phi/\partial X_\mu \tag{3.58}$$

## Second Quantized Coordinates

At this point we have developed a formalism for a scalar particle quantum field theory based on our non-commutative coordinates. In the following chapters we will proceed to use this formalism to develop a unified quantum field theory of the known forces of nature.

# 4. Scalar Two-Tier Quantum Field Theory

*It appears then that there must be fundamental changes in our basic formulation of quantum field theory, so that unrenormalized masses and unrenormalized coupling constants can become finite.*
*T. D. Lee & G. C. Wick*[17]

## Introduction

In this chapter we will examine a new formulation of quantum field theory that we call *two-tier quantum field theory* in more detail for the case of a free scalar particle. This type of quantum field theory incorporates a structure similar to a string-like substructure within a quantum field theoretic framework. In the following chapters we will apply the approach to QED, Electroweak Theory, the Standard Model and lastly to a unified model for the known forces of nature.

## "Two-Tier" Space

In the preceding chapter we developed quantized coordinates $X^\mu$ defined on an underlying c-number coordinates $y^\nu$ with the equations:

$$X_\mu(y) = y_\mu + iY_\mu(y)/M_c^2 \tag{3.12}$$

$$Y^i(y) = \int d^3k\, N_0(k) \sum_{\lambda=1}^{2} \varepsilon^i(k, \lambda)[a(k,\lambda)\, e^{-ik\cdot y} + a^\dagger(k,\lambda)\, e^{ik\cdot y}] \tag{3.24}$$

We also developed a free scalar quantum field theory with the Fourier expansion:

$$\phi(X) = \int d^3p\, N_m(p)\, [a(p)\, e^{-ip\cdot X} + a^\dagger(p)\, e^{ip\cdot X}] \tag{3.5}$$

---

[17] T. D. Lee and G. C. Wick, Phys. Rev. **D2**, 1033 (1970). Lee and Wick's model QED is totally unrelated to our approach.

We will now consider the implications of the separable Lagrangian:

$$\mathscr{L}_s = \mathscr{L}_F(\phi(X), \partial\phi/\partial X^\mu)\, J + \mathscr{L}_C(X^\mu(y), \partial X^\mu(y)/\partial y^\nu) \qquad (A.96)$$

where

$$\mathscr{L}_F = ½\,[\,(\partial\phi/\partial X^\nu)^2 - m^2\phi^2\,] \qquad (A.33)$$

and

$$\mathscr{L}_C = -¼\, M_c^{\,4} F_Y^{\ \mu\nu} F_{Y\mu\nu} \qquad (3.10)$$

with

$$F_{Y\mu\nu} = \partial Y_\mu/\partial y^\nu - \partial Y_\nu/\partial y^\mu \qquad (3.14)$$

$M_c$ is the mass that sets the scale at which the imaginary part of $X^\mu$ becomes significant.

This quantum field theory behaves as a conventional quantum field theory until energies reach the magnitude of $M_c$. At energies of the order of $M_c$, and above, the imaginary part of $X^\mu$ becomes significant and alters the high-energy behavior of the theory in a major way. This modification leads to the elimination of divergences that normally appear in perturbation theory when interactions are introduced. Yet the low energy behavior of the theory remains the same remains the same as conventional scalar quantum field theory. Thus the precise calculations of QED that have been verified to an amazing degree of accuracy remain valid when a two-tier formulation of QED is created (in chapter 5). And the "low energy" results found in other conventional quantum field theories such as Electroweak Theory and the Standard Model also are closely approximated by their corresponding two-tier versions.

The straightforward use of the above equations[18] (and the canonical quantization described in the preceding chapters) leads to a scalar quantum field with the Fourier expansion:

$$\phi(X) = \int d^3p\, N_m(p)\, [a(p)e^{-ip\cdot(y + iY/M_c^{\,2})} + a^\dagger(p)e^{ip\cdot(y + iY/M_c^{\,2})}] \qquad (4.1)$$

using eq. 3.5 above. We note the equal time commutation relations of $\phi$ and $\pi_\phi$ are the same as the conventional equal time commutation relations of a scalar field despite the fact that $X^\mu$ and $Y^\mu$ are themselves quantum fields since $[Y^\mu(\mathbf{y}, y^0), Y^\nu(\mathbf{y}', y^0)] = 0$ for $\mathbf{y} \neq \mathbf{y}'$. In addition, we note the $\phi$ and $\pi_\phi$ fields are not hermitean.

The Fourier expansion of $\phi$ does require one refinement – the exponential terms in $X^\mu$ must be *normal ordered* to avoid infinities in the unequal time commutation relations:

---

[18] The use of functionals in quantum field theory is, of course, far from new as one can see in texts such as Bogoliubov (1959) (see for example pp. 198-226).

$$\phi(X) = \int d^3p\, N_m(p)\, [a(p) :e^{-ip\cdot(y + iY/M_c^2)}: + a^\dagger(p) :e^{ip\cdot(y + iY/M_c^2)}:] \qquad (4.2)$$

Since the hamiltonian as well as other quantities are normal ordered in quantum field theory the additional requirement of normal ordering in the field operator is merely an extension of a standard procedure to a more complex situation and is not disturbing. The unequal time commutation relation of the normal ordered $\phi$ field is:

$$[\phi(X^\mu(y_1)), \phi(X^\mu(y_2))] = i\Delta(y_1 - y_2) + \mathcal{O}(1/M_c^2) \qquad (4.3)$$

where

$$\Delta(y_1 - y_2) = -i \int d^3k\, (e^{-ik\cdot(y_1 - y_2)} - e^{ik\cdot(y_1 - y_2)})/[(2\pi)^3 2\omega_k] \qquad (4.4)$$

is a familiar c-number invariant singular function. The additional terms in eq. 4.3 are q-number terms that become significant at very short distances of the order $M_c^{-1}$. Thus precise measurements of field strengths at larger distances are limited by standard quantum effects as indicated by the commutation relation.

The principle of *microscopic causality* is violated at extremely short distances of the order $M_c^{-1}$ since the commutator (eq. 4.3) is non-zero, in general, for space-like distances of the order of $M_c^{-1}$ due to the q-number terms. This violation is not experimentally measurable now – and for the foreseeable future – and reflects a type of non-locality at extremely short distances.

We will see that the short distance behavior of two-tier quantum field theory leads to the elimination of divergences resulting in finite interacting quantum field theories.

## Vacuum Fluctuations

While the expectation value of a *conventional* free scalar field $\phi_{conv}(X)$ is zero in a conventional quantum field theory:

$$<0|\phi_{conv}(X)|0> = 0 \qquad (4.5)$$

the vacuum fluctuations of *conventional* scalar quantum field theory are quadratically divergent:

$$<0|\phi_{conv}(X)\phi_{conv}(X)|0> = \int d^3p/[(2\pi)^3 2\omega_p] \qquad (4.6)$$

*In "two-tier" quantum field theory* we find the vacuum expectation value of a free field is zero (like eq. 4.5) *and the expectation value of the square of the field is also zero:*

$$<0|\phi(X)\phi(X)|0> = \int d^3p\ e^{-p^ip^j\Delta_{Tij}(0)/M_c^4}/[(2\pi)^3 2\omega_p] = 0 \quad (4.7)$$

since the exponential factor in the integral is $-\infty$. The exponent contains

$$\Delta_{Tij}(z) = \int d^3k\ e^{-ik\cdot z}\ (\delta_{ij} - k_ik_j/\mathbf{k}^2)/[(2\pi)^3 2\omega_k] \quad (4.8)$$

where "T"is for "Two-Tier". Thus *vacuum fluctuations are zero in two-tier quantum field theory*. Correspondingly, we will see that renormalization constants are finite in the two-tier versions of QED, Electroweak Theory, the Standard Model and Quantum Gravity.

## The Feynman Propagator

The Feynman propagator for a two-tier free scalar quantum field is:

$$i\Delta_F^{TT}(y_1 - y_2) = <0|T(\phi(X(y_1)),\phi(X(y_2)))|0> \quad (4.9)$$

$$\equiv <0|\phi(X(y_1))\phi(X(y_2))|0>\ \theta(y_1^0 - y_2^0) +$$

$$+\ \phi(X(y_2))\phi(X(y_1))|0>\ \theta(y_2^0 - y_1^0) \quad (4.10)$$

Since $X^0 = y^0$ in the Coulomb gauge of the $X^\mu$ field there is no ambiguity in the choice of the relevant time variable. A straightforward calculation shows:

$$i\Delta_F^{TT}(y_1 - y_2) = i\int d^4p\ e^{-ip\cdot(y_1 - y_2)}\ R(\mathbf{p}, y_1 - y_2)/[(2\pi)^4(p^2 - m^2 + i\varepsilon)] \quad (4.11)$$

where

$$R(\mathbf{p}, y_1 - y_2) = \exp[-p^ip^j\Delta_{Tij}(y_1 - y_2)/M_c^4] \quad (4.12)$$

$$= \exp\{-p^2[A(v) + B(v)\cos^2\theta] / [4\pi^2 M_c^4 z^2]\} \quad (4.13)$$

with

$$z^\mu = y_1^\mu - y_2^\mu \quad (4.14)$$

$$z = |\mathbf{z}| = |\mathbf{y}_1 - \mathbf{y}_2| \tag{4.15}$$

$$p = |\mathbf{p}| \tag{4.16}$$

$$v = |z^0|/z \tag{4.17}$$

$$A(v) = (1 - v^2)^{-1} + .5v \ln[(v-1)/(v+1)] \tag{4.18}$$

$$B(v) = v^2(1 - v^2)^{-1} - 1.5v \ln[(v-1)/(v+1)] \tag{4.19}$$

$$\mathbf{p}\cdot\mathbf{z} = pz\cos\theta \tag{4.20}$$

and $|\mathbf{p}|$ denoting the length of a spatial vector $\mathbf{p}$ while $|z^0|$ is the absolute value of $z^0$.

As eq. 4.11 indicates, the Gaussian damping factor R(p, z) for large momentum p is the same for both the positive and negative frequency parts of the two-tier Feynman propagator. It is also important to note that R(p, z) does not depend on $p^0$ (in the Y Coulomb gauge) and thus the integration over $p^0$ proceeds in the usual way to produce time-ordered positive and negative frequency parts.

*Large Distance Behavior of Two-Tier Theories*

The large distance behavior of the two-tier Feynman propagator approaches the behavior of the conventional Feynman propagator since

$$R(\mathbf{p}, y_1 - y_2) \to 1 \tag{4.21}$$

when $(y_1 - y_2)^2$ becomes much larger than $M_c^{-2}$ as eq. 4.13 shows. Thus the behavior of a conventional quantum field theory naturally emerges at large distance. We will see that the conventional Standard Model is the large distance limit of the two-tier Standard Model thus *realizing a form of Correspondence Principle for Quantum Field Theory.* Some features of the conventional Standard Model that depend specifically on the existence of divergences, such as the axial anomaly, will be different in the two-tier Standard Model since it is a divergence-free theory.

*Short Distance Behavior of Two-Tier Theories*

At short distances the Gaussian factor dominates and radically changes the behavior of the Feynman propagator eliminating its short distance singular behavior, and thus paving the way to finite quantum field theories. Near the light cone, $M_c^{-2} \gg -(y_1 - y_2)^2 \to 0$, we can approximate eq. 4.11 with

$$i\Delta_F^{TT}(y_1 - y_2) \approx \int d^3p\, [N(p)]^2\, R(\mathbf{p}, y_1 - y_2) \qquad (4.22)$$

since $e^{-ip\cdot(y_1 - y_2)}$ is approximately unity for small $(y_1 - y_2)$. We assume the mass of the $\phi$ particle is zero or is negligible at high energies so we set $m = 0$ to study the high energy behavior of eq. 4.22. Upon performing the integrations in eq. 4.22 for space-like $(y_1 - y_2)^2$ (and analytically continuing to the time-like regions[19,20]) we find

$$i\Delta_F^{TT}(y_1 - y_2) \approx [z^2M_c^4/(4i\surd A\surd B)]\ \ln[(\surd A + i\surd B)/(\surd A - i\surd B)] \qquad (4.23)$$

with A and B defined in eqs. 4.18 and 4.19. As $(y_1 - y_2)^2 \rightarrow 0$ from the space-like or time-like side of the light cone we find eq. 4.23 becomes:

$$i\Delta_F^{TT}(y_1 - y_2) \rightarrow \pi M_c^4\,|\,(y_1 - y_2)^\mu(y_1 - y_2)_\mu\,|\,/8 \qquad (4.24)$$

Eq. 4.24 has several noteworthy points:

1. The propagator is well behaved on the light cone and approaches zero smoothly from both space-like and time-like directions. In contrast, the conventional scalar Feynman propagator diverges as $[(y_1 - y_2)^\mu(y_1 - y_2)_\mu]^{-2}$. This good behavior near the light cone will be seen later for other particle propagators with the net result that the usual infinities found in conventional quantum field theory are absent in two-tier quantum field theories.

2. The quadratic form of the propagator in eq. 4.24 is suggestive of attempts to formulate a relativistic harmonic oscillator model of elementary particles[21] and more recent attempts to achieve quark confinement. The fact that the absolute value of the quadratic term appears in eq. 4.24 neatly avoids the common pitfall seen in fully relativistic harmonic oscillator attempts.

3. The quadratic behavior *in coordinate space* of the propagator at short distances is equivalent to a high-energy behavior of

$$p^{-6} \qquad (4.25)$$

---

[19] See S. Blaha, "Relativistic Bound State Models with Quasi-Free Constituent Motion", Phys. Rev. **D12**, 3921 (1975) and references therein.

[20] It should be noted that A and B in eq. 4.23 have the same sign for $0 \le v < 1.1243$ thus making for easy analytic continuation across the light cone (which corresponds to $v = 1$ in eqs. 4.18 and 4.19).

[21] H. Yukawa, H., Phys. Rev. **91**, 416 (1953); Y. S. Kim and M. E. Noz, Phys. Rev. **D8**, 3521 (1973) and references therein.

*in momentum space.* Thus we get the equivalent *of a higher derivative theory* in two-tier quantum field theory at high energies while retaining a positive definite energy spectrum. The problems of negative metric states that have plagued conventional higher derivative quantum field theories are avoided.[22]

## String-like Substructure of the Theory

Imaginary Quantum Dimensions™ endow a particle with an extended structure that resembles to some extent the extended structure seen in bosonic string and Superstring theories. For example, Bailin (1994) use the operator[23]

$$V_\Lambda(k) = \int d^2\sigma \sqrt{-h}\, W_\Lambda(\tau, \sigma)\, e^{-ik\cdot X} \qquad (4.26)$$

where $X^\mu$ is a quantized fourier expansion of the string fields (see eq. 7.22 of Bailin (1994)).

We note our $X^\mu$ coordinate-field has two transverse degrees of freedom due to gauge invariance, which also invites comparison to the bosonic string. A point of difference is that we will create a well-defined quantum field theoretic formulation in conventional space-time that has the Standard Model as its "large distance" behavior thus introducing a note of reality that is not (yet?) very apparent in Superstring theories. We see that the interacting quantum field theories based on this approach also have good, finite, short distance behavior just as string theories.

The scalar, and other particles', Feynman propagators can be viewed as describing the propagation of a particle cloaked (accompanied) by a cloud of Y particles (which generates the $R(\mathbf{p}, y_1 - y_2)$ factor in the propagator of eq. 4.11). If we examine the fourier transform of R(p, z) we see:

$$(2\pi)^4 R(\mathbf{p}, q) = \int d^4z\, e^{iq\cdot z} R(\mathbf{p}, z) = \int d^4z\, e^{iq\cdot z} \exp[-p^i p^j \Delta_{Tij}(z)/M_c^4] \qquad (4.27)$$

and we find

$$R(\mathbf{p},q) = \sum_{n=0}^{\infty} [i(2\pi M_c)^4]^{-n} (n!)^{-1} \prod_{j=1}^{n} [\int d^4k_j\, \theta(k_j^0)(\mathbf{p}^2 - (\mathbf{p}\cdot\mathbf{k}_j)^2/\mathbf{k}_j^2)/(k_j^2 + i\varepsilon)]\, \delta^4(q - \sum_r k_r) \qquad (4.28)$$

---

[22] S. Blaha, Phys.Rev. **D10**, 4268 (1974); S. Blaha, Phys.Rev. **D11**, 2921 (1975); S. Blaha, Nuovo Cim. **A49**, :113 (1979); S. Blaha, "Generalization of Weyl's Unified Theory to Encompass a Non-Abelian Internal Symmetry Group" SLAC-PUB-1799, Aug 1976; S. Blaha, "Quantum Gravity and Quark Confinement" Lett. Nuovo Cim. **18**, 60 (1977); Nakanishi, N., Suppl. Prog. Theo. Phys. **51**, 1 (1972); and references therein.

[23] D. Bailin and A. Love, *Supersymmetric Gauge Field Theory and String Theory* (Institute of Physics Publishing, Philadelphia, PA, 1994) page 272.

which can be interpreted as a "cloud" of Y particles dressing the "bare" particle propagator. (The manifest divergences in eq. 4.28 for R(p, q) are an artifact of the expansion and the subsequent fourier transformation. They are not present in the R(**p**, $y_1 - y_2$) factor in the propagator of eq. 4.11.) See Fig. 4.1 for the Feynman diagram of the two-tier cloaked propagator as compared to the normal scalar particle Feynman propagator. The two-tier Feynman propagator is basically a conventional scalar propagator that is modified by coherent Y particle emission.[24]

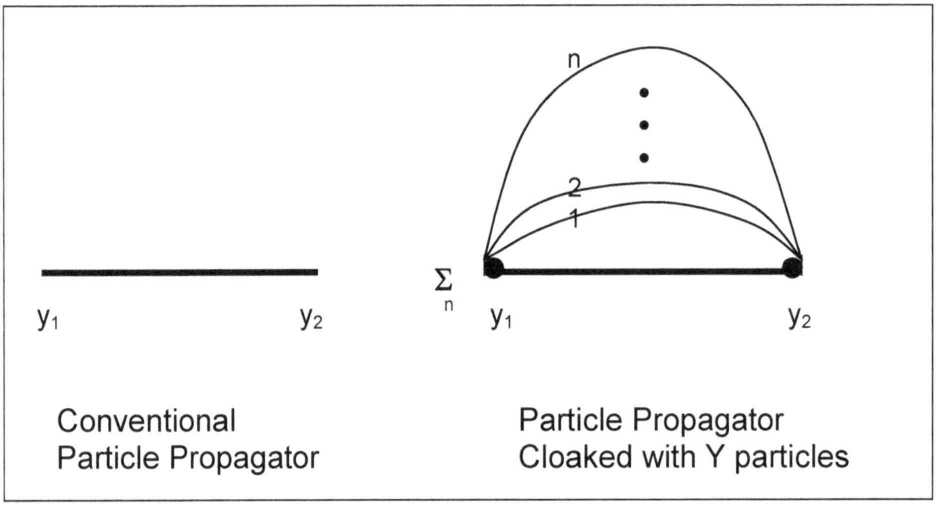


Figure 4.1. Feynman diagram for conventional and cloaked two-tier propagators.

We note that R(p, q) satisfies the convolution theorem:

$$\int d^4k\ R(\mathbf{p}, k)\ R(\mathbf{p}, q - k) = [R(\mathbf{p}, q)]^2 \tag{4.29a}$$

or

$$(2\pi)^4 \int d^4z\ e^{iq\cdot z}\ R(\mathbf{p}, z)\ R(\mathbf{p}, z) = [\ \int d^4z\ e^{iq\cdot z}\ R(\mathbf{p}, z)\ ]^2 \tag{4.29b}$$

The proof follows from eq. 4.28 and the Binomial theorem.

[24] T. W. B. Kibble, Phys. Rev. **173**, 1527 (1968) and references therein. In particular see p. 1532 of Kibble's paper.

## Parity

Parity can appear in two guises within the framework of two-tier quantum field theory. One can consider a parity operation where the space parts of $X^\mu$ are reversed while $y^\mu$ is unchanged. Or one can consider a second type of parity where the space parts of $y^\mu$ are reversed.

### *X Parity*

Under this form of parity operation $y^\mu$ is unchanged while the arguments of $\phi$ *appear* to change by

$$X^i(y) \to -X^i(y) \tag{4.30}$$

$$X^0(y) \to X^0(y) \tag{4.31}$$

We will denote the parity operator of this type $\mathscr{P}_X$. Under $\mathscr{P}_X$ the arguments of the scalar quantum field operator $\phi$ change according to eqs. 4.30-1 so that $\phi$ transforms as

$$\mathscr{P}_X\phi(\mathbf{X}(y), X^0(y))\mathscr{P}_X^{-1} = \phi(-\mathbf{X}(y), X^0(y)) \tag{4.32}$$

From the form of $\phi$ in eq. 4.2 we see we can implement eq. 4.32 by requiring:

$$\mathscr{P}_X a(\mathbf{p}, p^0)\mathscr{P}_X^{-1} = a(-\mathbf{p}, p^0) \tag{4.33}$$

$$\mathscr{P}_X a^\dagger(\mathbf{p}, p^0)\mathscr{P}_X^{-1} = a^\dagger(-\mathbf{p}, p^0) \tag{4.34}$$

$$\mathscr{P}_X X^0(y)\mathscr{P}_X^{-1} = X^0(y) \tag{4.35}$$

$$\mathscr{P}_X X^i(y)\mathscr{P}_X^{-1} = X^i(y) \tag{4.36}$$

$$\mathscr{P}_X Y^i(y)\mathscr{P}_X^{-1} = Y^i(y) \tag{4.37}$$

where i = 1,2,3.

This parity transformation is analogous to the standard form of parity transformation in conventional quantum field theory. The separable Lagrangian in eq. A.96 (and listed at the beginning of this chapter) is invariant under this parity transformation.

*y Parity*

This form of parity transformation in which $y^i \to -y^i$ has significant differences from the normal parity transformation. We specify this parity transformation for a scalar quantum field by:

$$\mathscr{P}_y \phi(\mathbf{X}(\mathbf{y}, y^0), X^0(\mathbf{y}, y^0))\mathscr{P}_y^{-1} = \phi(\mathbf{X}(-\mathbf{y}, y^0), X^0(-\mathbf{y}, y^0)) \tag{4.38}$$

This transformation can be implemented through the following set of transformations:

$$\mathscr{P}_y a(\mathbf{p}, p^0)\mathscr{P}_y^{-1} = a(-\mathbf{p}, p^0) \tag{4.39}$$

$$\mathscr{P}_y a^\dagger(\mathbf{p}, p^0)\mathscr{P}_y^{-1} = a^\dagger(-\mathbf{p}, p^0) \tag{4.40}$$

$$\mathscr{P}_y X^0(\mathbf{y}, y^0)\mathscr{P}_y^{-1} = X^0(-\mathbf{y}, y^0) \tag{4.41}$$

$$\mathscr{P}_y Y^i(\mathbf{y}, y^0)\mathscr{P}_y^{-1} = -Y^i(-\mathbf{y}, y^0) \tag{4.42a}$$

$$\mathscr{P}_y a(\mathbf{k}, k^0,1)\mathscr{P}_y^{-1} = a(-\mathbf{k}, k^0,1) \tag{4.42b}$$

$$\mathscr{P}_y a(\mathbf{k}, k^0,2)\mathscr{P}_y^{-1} = -a(-\mathbf{k}, k^0,2) \tag{4.42c}$$

where i = 1,2,3 and assuming: $\boldsymbol{\varepsilon}(\mathbf{k},k^0,1)=-\boldsymbol{\varepsilon}(-\mathbf{k},k^0,1)$ and $\boldsymbol{\varepsilon}(\mathbf{k},k^0,2)=+\boldsymbol{\varepsilon}(-\mathbf{k},k^0,2)$.

*Forms of the Parity Transformations*

The parity transformations for a scalar particle are

$$\mathscr{P}_X = \exp\{-i\pi\int d^3p\ [a^\dagger(\mathbf{p}, p^0)a(\mathbf{p}, p^0) - a^\dagger(\mathbf{p}, p^0)a(-\mathbf{p}, p^0)]/2\} \tag{4.43a}$$

$$\mathscr{P}_y = \mathscr{P}_X \exp\{-i\pi \int d^3k\ [\sum_{\lambda=1}^{2} a^\dagger(\mathbf{k},k^0,\lambda)a(\mathbf{k},k^0,\lambda) - a^\dagger(\mathbf{k}, k^0, 1)a(-\mathbf{k},k^0,1) + $$
$$+ a^\dagger(\mathbf{k}, k^0, 2)a(-\mathbf{k},k^0,2)]/2\} \tag{4.43b}$$

The separable Lagrangian of eq. A.96 is invariant under these parity transformations.

## Charge Conjugation

Charge conjugation is implemented in a way similar to that of conventional quantum field theory. In particular

$$\mathscr{C}X^{\mu}(\mathbf{y}, y^0)\mathscr{C}^{-1} = X^{\mu}(\mathbf{y}, y^0) \tag{4.44}$$

## Time Reversal

Since $X^0 = y^0$ in the Y Coulomb gauge in two-tier quantum theory the only non-trivial form of time reversal transformation $\mathscr{T}$ is based on $y^0 = -y^0$. This time reversal transformation is similar in part to to the conventional time reversal transformation in conventional quantum field theory. Therefore we will define $\mathscr{T}$ as the product of the operation of taking the complex conjugate of all c-numbers times a unitary operator $\mathscr{U}_Y$. Under $\mathscr{T}$ a scalar quantum field operator $\phi$ transforms as

$$\mathscr{T}\phi(\mathbf{X}(\mathbf{y}, y^0), X^0(\mathbf{y}, y^0))\mathscr{T}^{-1} = \phi(\mathbf{X}(\mathbf{y}, -y^0), X^0(\mathbf{y}, -y^0)) \tag{4.45}$$

From the form of in $\phi$ eq. 4.2 we see that

$$\mathscr{T}\, a(\mathbf{p}, p^0)\mathscr{T}^{-1} = a(-\mathbf{p}, p^0) \tag{4.46}$$

$$\mathscr{T}\, a^{\dagger}(\mathbf{p}, p^0)\mathscr{T}^{-1} = a^{\dagger}(-\mathbf{p}, p^0) \tag{4.47}$$

$$\mathscr{T}\, X^i(\mathbf{y}, y^0)\mathscr{T}^{-1} = X^i(\mathbf{y}, -y^0) \tag{4.48}$$

$$\mathscr{T}\, Y^i(\mathbf{y}, y^0)\mathscr{T}^{-1} = -Y^i(\mathbf{y}, -y^0) \tag{4.49a}$$

$$\mathscr{T}\, a(\mathbf{k}, k^0,1)\mathscr{T}^{-1} = a(-\mathbf{k}, k^0,1) \tag{4.49b}$$

$$\mathscr{T}\, a(\mathbf{k}, k^0,2)\mathscr{T}^{-1} = -a(-\mathbf{k}, k^0,2) \tag{4.49c}$$

where i = 1,2,3 and assuming: $\boldsymbol{\varepsilon}(\mathbf{k},k^0,1)=-\boldsymbol{\varepsilon}(-\mathbf{k},k^0,1)$ and $\boldsymbol{\varepsilon}(\mathbf{k},k^0,2)=+\boldsymbol{\varepsilon}(-\mathbf{k},k^0,2)$.

The unitary operator $\mathscr{U}_Y$ is given by

$$\mathscr{U}_X = \exp\{-i\pi\int d^3p\ [a^\dagger(\mathbf{p}, p^0)a(\mathbf{p}, p^0) - a^\dagger(\mathbf{p}, p^0)a(-\mathbf{p}, p^0)]/2\} \qquad (4.50a)$$

and

$$\mathscr{U}_y = \mathscr{U}_X \exp\{-i\pi \int d^3k\ [\sum_{\lambda=1}^{2} a^\dagger(\mathbf{k},k^0,\lambda)a(\mathbf{k},k^0,\lambda) - a^\dagger(\mathbf{k}, k^0, 1)a(-\mathbf{k},k^0,1) +$$
$$+ a^\dagger(\mathbf{k}, k^0, 2)a(-\mathbf{k},k^0,2)]/2\} \qquad (4.50b)$$

The separable Klein-Gordon Lagrangian (eq. A.96) is invariant under our definition of time reversal.

We note

$$\mathscr{U}_y = \mathscr{P}_y \qquad (4.50c)$$

Although the present theory is somewhat more complicated than conventional quantum field theory the overall nature of the $\mathscr{P}$, $\mathscr{C}$, and $\mathscr{T}$ transformations is the same.

# 5. Interacting Quantum Field Theory – Perturbation Theory

## Introduction

The form of quantum field theory that we have developed in chapters 3 and 4 can be used as the basis for new formulations of QED, Electroweak Theory, the Standard Model and a divergence-free, unified theory of all the known interactions. The development of these theories requires a number of topics be addressed. This chapter covers perturbation theory. As much as possible, we attempt to retain the features of the standard approach so that the reader will more readily follow the discussion and more readily accept this new formalism. In physics originality is secondary to reality. The perturbation theory that we will develop will be shown to be identical to the perturbation theory that we develop later using a path integral formalism.

*Two-tier theory will be shown to satisfy unitarity in chapter 6 and invariant under special relativity in Appendix B.*

## An Auxiliary Asymptotic Field

The definition of the asymptotic "free" in and out states is an issue in two-tier quantum field theory because the "free particle field" of the theory $\phi(X(y))$ is a "dressed" particle, ab initio, since it is cloaked in a cloud of Y particles as discussed in the passage following eq. 4.27.

While one could use $\phi(X(y))$ directly to define in and out asymptotic states it is more convenient initially to introduce a "fictitious" auxiliary asymptotic quantum field $\Phi(y)$ that will represent the equally fictitious "bare $\phi$ particle" in and out states.

We will consider the case of a scalar field. We define a free, scalar Klein-Gordon particle field with the physical mass m of the physical $\phi(X(y))$ particle.

$$\Phi(y) = \int d^3p\, N_m(p)\, [a(p)\, e^{-ip\cdot y} + a^\dagger(p)\, e^{ip\cdot y}] \tag{5.1}$$

using the creation and annihilation operators of $\phi(X(y))$ (in eq. 4.2). The set of particle states of $\Phi(y)$ has the familiar Fock space form

$$| p_1, p_2, \ldots p_n\rangle = a^\dagger(p_1)a^\dagger(p_2) \ldots a^\dagger(p_n)|0\rangle \tag{5.2}$$

with powers of creation operators allowed since Φ particles are bosons. The set of particle states constitutes a complete orthonormal set of states. The corresponding bra states are defined by hermitean conjugation:

$$\langle p_1, p_2, \ldots p_n| = (| p_1, p_2, \ldots p_n\rangle)^\dagger \tag{5.3}$$

We note that the energy spectrum of these states is positive definite with the hamiltonian

$$H_\Phi = P_\Phi{}^0 = \int d^3y \ {}^1\!/\!_2 \, [\pi_\Phi{}^2 + (\partial\Phi/\partial X^i)^2 + m^2\Phi^2] \tag{5.4a}$$

$$= \int d^3p \ (\mathbf{p}^2 + m^2)^{1/2} a^\dagger(p)a(p) \tag{5.4b}$$

and momentum vector:

$$\mathbf{P}_\Phi = \int d^3p \ \mathbf{p} \ a^\dagger(p)a(p) \tag{5.5}$$

We will use this set of energy-momentum eigenstates to define asymptotic “in” and “out” states in perturbation theory.

## Transformation Between Φ(y) and $\phi$(X(y))

For later use in the definition of the perturbation theory expansion, we will determine the transformations between the in and out Φ(y) fields, and the in and out $\phi(X(y))$ fields. Let us define a transformation $W_a(y)$ that transforms in and out Φ(y) fields to in and out $\phi(X(y))$ fields respectively:

$$\phi_a(X(y)) = :W_a(y)\Phi_a(y)W_a^{-1}(y): \tag{5.6}$$

where the label a = “in” or a = “out”, where : ... : signifies normal ordering, and where

$$\Phi_{in}(y) = \int d^3p \ N_m(p) \ [a_{in}(p) \ e^{-ip\cdot y} + a_{in}{}^\dagger(p) \ e^{ip\cdot y}] \tag{5.7}$$

$$\Phi_{out}(y) = \int d^3p \ N_m(p) \ [a_{out}(p) \ e^{-ip\cdot y} + a_{out}{}^\dagger(p) \ e^{ip\cdot y}] \tag{5.8}$$

$$\phi_{in}(X) = \int d^3p\, N_m(p)\, [a_{in}(p)\, :e^{-ip\cdot(y + iY/M_c^2)}: + a_{in}^\dagger(p)\, :e^{ip\cdot(y + iY/M_c^2)}:] \quad (5.9)$$

$$\phi_{out}(X) = \int d^3p\, N_m(p)\, [a_{out}(p)\, :e^{-ip\cdot(y + iY/M_c^2)}: + a_{out}^\dagger(p)\, :e^{ip\cdot(y + iY/M_c^2)}:] \quad (5.10)$$

Note that the transformation eq. 5.6 includes normal ordering. While this transformation may seem strange it is no stranger than the time reversal operator, in which the complex conjugate of all c-number terms is taken in addition to applying a unitary transformation.

In the Coulomb gauge of Y it is easy to show that

$$W_a(y) = \exp(-\mathbf{Y}(y)\cdot\mathbf{P}_{\Phi a}/M_c^2) \quad (5.11)$$

and

$$W_a^{-1}(y) = \exp(\mathbf{Y}(y)\cdot\mathbf{P}_{\Phi a}/M_c^2) \quad (5.12)$$

where the label a = "in" or a = "out", where the inner products in the exponentials are the usual spatial vector inner product, and where

$$\mathbf{P}_{\Phi a} = -\int d^3y\, \partial\Phi_a(y)/\partial y^0\, \nabla\Phi_a(y) = \int d^3p\, \mathbf{p}\, a_a^\dagger(p)a_a(p) \quad (5.12a)$$

is a spatial vector (the Φ spatial momentum operator) that is written solely in terms of $\Phi_a(y)$'s creation and annihilation operators.

In addition to performing the transformation in eq. 5.6 $W_a(y)$ also performs a "translation" in $Y^\mu$:

$$W_a(y)Y^i(y')W_a^{-1}(y) = Y^i(y') + i\Delta^{trij}(y' - y)P_{\Phi a}^j/M_c^2 \quad (5.13a)$$

where

$$i\Delta^{trij}(y' - y) = \int d^3k\, (e^{-ik\cdot(y' - y)} - e^{ik\cdot(y' - y)})(\delta_{jk} - k_jk_k/\mathbf{k}^2)/[(2\pi)^3 2\omega_k] \quad (5.13b)$$

We note that $W_a(y)$ is not a unitary operator but it is pseudo-unitary:

$$W_a^{-1}(y) = V\, W_a^\dagger(y)\, V^{-1} = V\, W_a(y)\, V^{-1} \quad (5.14)$$

where

$$V = \exp(-i\pi \sum_{\lambda=1}^{2} \int d^3k\, a^\dagger(k, \lambda)a(k, \lambda)) \quad (5.15)$$

is a unitary operator with the property

$$V\,Y^j(y)\,V^{-1} = -Y^j(y) \tag{5.16}$$

for j = 1,2,3. We note

$$V^\dagger = V^{-1} = V \tag{5.17}$$

and thus

$$V^2 = I \tag{5.18}$$

V will be shown to be a metric operator in the following discussion.[25] We note that the Y "particle" (hermitean) number operator appears in eq. 5.9 in the expression for V:

$$N_Y = \sum_{\lambda=1}^{2} \int d^3k\, a^\dagger(k, \lambda) a(k, \lambda) \tag{5.19}$$

Thus states with an even number of Y "particles" have a V eigenvalue of one, and states with an odd number of Y "particles" have a V eigenvalue of minus one.

## Model Lagrangian with $\phi^4$ Interaction

We will develop our perturbation theory using a scalar Lagrangian model with a $\phi^4$ interaction term:

$$\mathscr{L}_s = J\mathscr{L}_F + \mathscr{L}_C \tag{5.20}$$

with

$$\mathscr{L}_F = ½\,[\,(\partial\phi/\partial X^\nu)^2 - m^2\phi^2] + \mathscr{L}_{Fint} \tag{5.21}$$

and

$$\mathscr{L}_C = -¼\,F_Y^{\mu\nu}F_{Y\mu\nu} \tag{5.22}$$

with

$$F_{Y\mu\nu} = \partial Y_\mu/\partial y^\nu - \partial Y_\nu/\partial y^\mu \tag{5.23}$$

and

$$\mathscr{L}_{Fint} = 1/4!\, \chi_0\, \phi(X(y))^4 + ½\,(m^2 - m_0^2)\phi^2 \tag{5.24}$$

where J is the Jacobian (as in Appendix A), $\chi_0$ is the bare coupling constant, and $m_0$ is the bare mass.

---

25 P. A. M. Dirac, Proc. R. Soc. London A **180**, 1 (1942); T. D. Lee and G. C. Wick, Nucl. Phys. **B9**, 209 (1969); C. M. Bender, D. C. Brody and H. F. Jones, "Complex Extension of Quantum Mechanics" Phys. Rev. Letters **89**, 270401-1 (2002) and references therein.

The conserved momentum operator is:

$$\mathsf{P}_{F\beta} = \int d^3X \, \mathscr{T}_{F0\beta} \tag{5.25}$$

where

$$\mathscr{T}_{F\mu\nu} = - g_{\mu\nu} \mathscr{L}_F + \partial \mathscr{L}_F \,/\, \partial(\partial\phi/\partial X_\mu) \, \partial\phi/\partial X^\nu \tag{5.26}$$

is the $\phi$ field energy-momentum tensor with conservation law (eq. A.110):

$$\partial P_{F\beta}/\partial X^0 = 0 \tag{5.27}$$

due to eq. A.108.

The hamiltonian density (eq. A.83) is

$$\mathscr{H}_F = \mathscr{T}_{F0\beta} = \mathscr{H}_{F0} + \mathscr{H}_{Fint} \tag{5.28}$$

with

$$\mathscr{H}_{F0} = \tfrac{1}{2} \, [\pi_\phi^{\,2} + (\partial\phi/\partial X^i)^2 + m^2\phi^2] \tag{5.29}$$

$$\mathscr{H}_{Fint} = - \tfrac{1}{4!} \, \chi_0 \, \phi(X(y))^4 + \tfrac{1}{2} \, (m^2 - m_0^{\,2})\phi(X(y))^2 \tag{5.30}$$

## In-states and Out-States

In this section we will develop properties of in-fields and out-fields. We will use a somewhat more complicated procedure to set up the perturbation theory for the S matrix due to the introduction of imaginary coordinates. The procedure can be schematized as:

$$\Phi_{in}(y) \Rightarrow \phi_{in}(X(y)) \Rightarrow \phi(X(y)) \Rightarrow \phi_{out}(X(y)) \Rightarrow \Phi_{out}(y) \tag{5.31}$$

In-states are constructed using the auxiliary field $\Phi_{in}$ which are then effectively transformed into $\phi_{in}(X(y))$ expressions in order to make contact with our Lagrangian formalism. Then $\phi_{in}(X(y))$ is related to the interacting field $\phi(X(y))$ as a limit ($y^0 \rightarrow -\infty$). Similarly out-states are constructed using the auxiliary field $\Phi_{out}$ which are then expressed in terms of $\phi_{out}(X(y))$. Then $\phi_{out}(X(y))$ is related to the interacting field $\phi(X(y))$ using the LSZ limiting process ($y^0 \rightarrow +\infty$).

Since much of the development differs only trivially from the standard treatment in textbooks we will simply "list" relevant equations and let the reader pursue them further in quantum field theory textbooks.

*$\phi$ In-Field*

In order to define a perturbation theory for particle scattering we will next specify features of the in-field $\phi_{in}(X(y))$ and in-field states – the field and states representing physical particles as $X^0 = y^0 \rightarrow -\infty$.

A. The in-field $\phi_{in}(X(y))$ satisfies the Klein-Gordon equation in the X variable:

$$(\Box_X + m^2)\,\phi_{in}(X) = 0 \qquad (5.32)$$

where

$$\Box_X = (\partial/\partial X^\nu)(\partial/\partial X_\nu)$$

B. Under coordinate displacements and Lorentz transformations $\Phi_{in}(y)$, $\phi_{in}(X(y))$, and $\phi(X(y))$ transform in the same way:

$$[P^\mu, \Phi_{in}(y)] = -\,i\partial\Phi_{in}/\partial y_\mu \qquad (5.33a)$$

$$[P^\mu, \phi_{in}(X)] = -\,i\partial\phi_{in}/\partial y_\mu \qquad (5.33b)$$

$$[P^\mu, \phi(X)] = -\,i\partial\phi/\partial y_\mu \qquad (5.34)$$

with the energy-momentum vector $P^\mu$ specified by eq. A.57.

C. We can relate the asymptotic in-field $\phi_{in}(X(y))$ to the interacting field $\phi(X(y))$ using the equation of motion of $\phi(X(y))$

$$(\Box_X + m^2)\,\phi(X) = j(X) \qquad (5.35)$$

where j(X) embodies the interaction. Using the physical mass m we find

$$(\Box_X + m^2)\,\phi(X) = j(X) + (m^2 - m_0^2)\phi(X) = j_{tot}(X) \qquad (5.36)$$

If the current is taken to be the source of the scattered waves we may write

$$\sqrt{Z}\,\phi_{in}(X(y)) = \phi(X(y)) - \int d^4X(y')\,\Delta_{ret}(y - y')\,j_{tot}(X(y')) \qquad (5.37)$$

$$= \phi(X(y)) - \int d^4y' \, J \, \Delta_{ret}(y - y') \, j_{tot}(X(y')) \tag{5.38}$$

where Z is a wave function renormalization constant, J is the Jacobian, and $\Delta_{ret}$ is a retarded Green's function.

D. We can define $\Phi_{in}$ in-field states with expressions like

$$| \, p_1, p_2, \ldots p_n \, in> = a_{in}^{\dagger}(p_1) a_{in}^{\dagger}(p_2) \ldots a_{in}^{\dagger}(p_n) \, | \, 0> \tag{5.39}$$

with powers of creation operators allowed since $\Phi_{in}$ is a boson field. The set of all particle states constitutes a complete orthonormal set of states. The corresponding bra states are defined by hermitean conjugation:

$$<p_1, p_2, \ldots p_n \, in \, | = (| \, p_1, p_2, \ldots p_n \, in>)^{\dagger} \tag{5.40}$$

*$\phi$ Out-Field*

In order to define a perturbation theory for particle scattering we begin by listing aspects of the out-field $\phi_{out}(X(y))$ and out-field states – the field and states representing physical particles as $X^0 = y^0 \rightarrow -\infty$.

A. The out-field $\phi_{out}(X(y))$ satisfies the Klein-Gordon equation in the X variable:

$$(\Box_X + m^2) \, \phi_{out}(X) = 0 \tag{5.41}$$

where

$$\Box_X = (\partial/\partial X^{\nu})(\partial/\partial X_{\nu})$$

B. Under coordinate displacements and Lorentz transformations $\Phi_{out}(y)$, $\phi_{out}(X(y))$, and $\phi(X(y))$ transform in the same way:

$$[P^{\mu}, \Phi_{out}(y)] = - i\partial\Phi_{out}/\partial y_{\mu} \tag{5.42a}$$

$$[P^{\mu}, \phi_{out}(X)] = - i\partial\phi_{out}/\partial y_{\mu} \tag{5.42b}$$

$$[P^{\mu}, \phi(X)] = - i\partial\phi/\partial y_{\mu} \tag{5.43}$$

with the energy-momentum vector $P^{\mu}$ specified by eq. A.57.

C. We can relate the asymptotic out-field $\phi_{out}(X(y))$ to the interacting field $\phi(X(y))$ using the equation of motion of $\phi(X(y))$ specified by eq. 5.36:

$$\sqrt{Z}\ \phi_{out}(X(y)) = \phi(X(y)) - \int d^4X(y')\ \Delta_{adv}(y - y')\ j_{tot}(X(y')) \tag{5.44}$$

$$= \phi(X(y)) - \int d^4y'\ J\ \Delta_{adv}(y - y')\ j_{tot}(X(y')) \tag{5.45}$$

where Z is a wave function renormalization constant, J is the Jacobian, and $\Delta_{adv}$ is an advanced Green's function.

D. We can define $\Phi_{out}$ out-field states with expressions like

$$|\ p_1, p_2, \ldots\ p_n\ out> = a_{out}^{\dagger}(p_1)a\Phi_{out}^{\dagger}(p_2)\ \ldots\ a\Phi_{out}^{\dagger}(p_n)\,|0> \tag{5.46}$$

with powers of creation operators allowed since $\Phi_{out}$ is a boson field. The set of all particle states constitutes a complete orthonormal set of states. The corresponding bra states are defined by hermitean conjugation:

$$<p_1, p_2, \ldots\ p_n\ out| = (|\ p_1, p_2, \ldots\ p_n\ out>)^{\dagger} \tag{5.47}$$

*The Y Field*

The Y field in the present model Lagrangian (eq. 5.20) is a free field and thus:

$$Y_{in}(y) = Y_{out}(y) = Y(y) \tag{5.48}$$

The states of the Y field have two general forms: 1) States in a Fock space consisting of particle states that are eigenstates of the Y particle number operator (eq. 5.19); and 2) Coherent states in a non-Fock space of generalized coherent states in an infinite tensor product space.[26]

The coherent ket states that arise in two-tier quantum field theory have the general form (eq. 3.41):

$$|y,\ p> = e^{-\mathbf{p}\cdot\mathbf{Y}^{-}(y)/M_c^2}\,|0> \tag{3.41}$$

as can be seen from an examination of $\phi_{in}(X(y))$. The corresponding bra state is:

---

[26] See Kibble and other references on coherent states.

$$<y, p| = (V| y, p>)^{\dagger} = <0|e^{+\mathbf{p}\cdot\mathbf{Y}^{+}(y)/M_c^2} \qquad (5.49)$$

with V, the metric operator, reversing the sign of Y in the exponential. The inner product of coherent states is:

$$<y_1, p_1| y_2, p_2> = \exp[-p_1^i p_2^j \Delta_{Tij}(y_1 - y_2)/M_c^4] \qquad (5.50)$$

showing the set of coherent states is not orthonormal and, in fact, is overcomplete. Comparing eq. 5.50 to eq. 4.12 gives

$$<y_1, p| y_2, p> = R(p, y_1 - y_2) \qquad (5.50a)$$

The completeness of the set of states for each time $y^0$ can be verified by examining the projection operator:

$$\mathscr{R}_Y(y^0) = \because \exp[-i \int d^3y\, Y^-_i(y)|0><0|\pi^{+i}(y)] \because \qquad (5.51)$$

where

$$\pi^{+j}(y) = -\partial Y^{+j}(y)/\partial y^0 \qquad (5.52)$$

and where $\because$ represents an extended normal ordering operator:

$$\because \ldots \because$$

which is defined as placing creation operators to the left, projection operators in the center, and annihilation operators to the right. Thus eq 5.51 can be written

$$\mathscr{R}_Y = \sum_n (-i/n!)^n \int d^3y_1 \ldots \int d^3y_n Y^{-j_1}(y_1)Y^{-j_2}(y_2)\ldots Y^{-j_n}(y_n)|0><0|\pi^+_{j_1}(y_1)\pi^+_{j_2}(y_2)\ldots\pi^+_{j_n}(y_n) \qquad (5.53)$$

where we have used the fact that $|0><0|$ is a projection operator, and reduced $|0><0|$ $|0><0|$ ... $|0><0|$ to $|0><0|$ in eq. 5.53. The vacuum state is the product of the Y and $\phi$ vacuum states:

$$|0> = |0_Y>|0_\phi> \qquad (5.53a)$$

We note

$$\mathscr{R}_Y(y^0)\,|\mathbf{y},y^0\,p> \;=\; |\mathbf{y},y^0\,p> \tag{5.54}$$

using eq. 3.22 and $\int d^3y_2\, p^i\, \Delta^{tr}{}_{ij}(y_1 - y_2)Y^{+j}(y_2) = \mathbf{p\cdot Y}^+(y_1)$. Also

$$\mathscr{R}_Y(y^0)\,|n> \;=\; |n> \tag{5.55}$$

where $|n>$ is any Y particle Fock state of finite particle number. In view of eqs. 5.54 and 5.55, we see that $\mathscr{R}_Y$ is the identity operator on the Fock space and the space of generalized coherent Y field states. Thus the set of Y coherent states forms an overcomplete set of states. We will define the S matrix for any combination of Φ Fock space states and coherent Y states. The $\mathscr{R}_Y$ operator can be generalized to include Φ Fock space states:

$$\mathscr{R}_{\Phi Y}(y^0) = \because \exp[-i\int d^3y\; Y^-{}_j(y)\mathscr{R}_\Phi\pi^{+j}(y)]\because \tag{5.56}$$

with

$$\mathscr{R}_\Phi = \sum_n\; |n><n| \tag{5.57}$$

is a sum over all Φ Fock space states with vacuum state given by eq. 5.53a. Since $\mathscr{R}_\Phi$ is a projection:

$$[\mathscr{R}_\Phi\,]^N = \mathscr{R}_\Phi$$

for any power N, we find:

$$\mathscr{R}_{\Phi Y}(y^0) = \sum_n\,(-i)^n\int d^3y_1\ldots\int d^3y_n Y^{-j_1}(y_1)Y^{-j_2}(y_2)\ldots Y^{-j_n}(y_n)\mathscr{R}_\Phi\pi^+{}_{j_1}(y_1)\pi^+{}_{j_2}(y_2)\ldots\pi^+{}_{j_n}(y_n) \tag{5.58}$$

As a result we have

$$\mathscr{R}_{\Phi Y}(y^0)\,|y,\,p;\,n_\Phi> \;=\; |y,\,p;\,n_\Phi> \tag{5.59}$$

for any combination of Y coherent states and Φ Fock space states $n_\Phi$. Also

$$\mathscr{R}_{\Phi Y}(y^0)\,|n_\Phi> \;=\; |n_\Phi> \tag{5.60}$$

Thus $\mathscr{R}_{\Phi Y}$ is the identity operator on this space – the (over) complete space of in and out states which we will use to define the S matrix of the scalar field theory specified by the Lagrangian eq. 5.20.

## S Matrix

Following the standard definition of the S matrix we have:

$$S_{\alpha\beta} = <\alpha \text{ out} | \beta \text{ in}> \tag{5.61}$$

$$= <\alpha \text{ in} | S | \beta \text{ in}> \tag{5.62}$$

$$|0> = |0 \text{ in}> = |0 \text{ out}> = S|0 \text{ in}> \tag{5.63}$$

$$\Phi_{in}(y) = S\Phi_{out}(y)S^{-1} \tag{5.64}$$

and the other standard properties of the S matrix with the sole exception being the form of the unitarity relation (discussed later).

## LSZ Reduction for Scalar Fields

In this section we will determine the reduction formula for the S matrix for scalar $\phi$ fields. Consider the S matrix element corresponding to an in state of particles $\beta$ plus one $\phi$ particle of momentum p, and an out state $\alpha$:

$$S_{\alpha\beta p} = <\alpha \text{ out} | \beta p \text{ in}> \tag{5.65}$$

After standard manipulations we have:

$$S_{\alpha\beta p} = <\alpha - p \text{ out} | \beta \text{ in}> - i<\alpha \text{ out} | \int d^3y \; f_p(y) \overleftrightarrow{\partial}_0 \, [\Phi_{in}(y) - \Phi_{out}(y)] \; |\beta \text{ in}> \tag{5.66}$$

where $<\alpha - p \text{ out}|$ is an out state with a particle of momentum p removed (if present) and where

$$f(y^0) \overleftrightarrow{\partial}_0 \, g(y^0) = f(y^0) \; \partial g(y^0)/\partial y^0 - \partial f(y^0)/\partial y^0 \; g(y^0) \tag{5.67}$$

and

$$f_p(y) = N_m(p)e^{-ip\cdot y} \tag{5.68}$$

with $N_m(p)$ specified by eq. 3.6.
We now express

$$S_{\alpha\beta p} = S_{\alpha-p\beta} - i\langle\alpha\ \text{out}|\int d^3y\ f_p(y)\ \overleftrightarrow{\partial}_0 W^{-1}[\phi_{in}(X(y)) - \phi_{out}(X(y))]W|\beta\ \text{in}\rangle \tag{5.69}$$

using $W(y) = W_{in}(y)$ with

$$\Phi_a(y) = W_a^{-1}(y)\phi_a(X(y))W_a(y) \tag{5.70}$$

where the label a = "in" or a = "out", and where

$$W_a(y) = \exp(-\mathbf{Y}(y)\cdot\mathbf{P}_{\Phi a}/M_c^2) \tag{5.71}$$

and

$$W_a^{-1}(y) = \exp(\mathbf{Y}(y)\cdot\mathbf{P}_{\Phi a}/M_c^2) \tag{5.72}$$

in the Coulomb gauge of Y with $\mathbf{P}_{\Phi a}$ the momentum spatial vector defined by eq. 5.12a.

We note that the interacting $\phi(X(y))$ approaches the in and out fields $\phi_{in}(X(y))$ and $\phi_{out}(X(y))$ in the limit that $y^0 \to -\infty$ and $y^0 \to +\infty$ respectively in the sense of Lehmann, Symanzik and Zimmermann[27] which we *symbolize* as:

$$\phi(X(y)) \to \sqrt{Z}\ \phi_{in}(X(y)) \qquad \text{as}\quad y^0 \to -\infty \tag{5.73}$$

$$\phi(X(y)) \to \sqrt{Z}\ \phi_{out}(X(y)) \qquad \text{as}\quad y^0 \to +\infty \tag{5.74}$$

with $\sqrt{Z}$ defined in eqs. 5.37 and 5.44. Thus we can rewrite eq. 5.69 as

$$S_{\alpha\beta p} = S_{\alpha-p\beta} + iZ^{-1/2}\left(\lim_{y_0\to+\infty} - \lim_{y_0\to-\infty}\right)\langle\alpha\ \text{out}|\int d^3y\ f_p(y)\ \overleftrightarrow{\partial}_0 W^{-1}\phi(X(y))W|\beta\ \text{in}\rangle \tag{5.75}$$

which after standard manipulations becomes

---

[27] H. Lehmann, K. Symanzik and W. Zimmermann, Nuov. Cim., 1, 1425 (1955); W. Zimmermann, Nuov. Cim., **10**, 567 (1958); O. W. Greenberg, Doctoral Dissertation, Princeton University 1956.

$$S_{\alpha\beta p} = S_{\alpha-p\beta} + iZ^{-1/2} \int d^4y\, f_p(y)(\Box_y + m^2)<\alpha\ \text{out}|\ W(y)^{-1}\phi(X(y))W(y)\,|\beta\ \text{in}> \tag{5.76}$$

Eq. 5.76 is similar to the usual LSZ reduction formula except for the appearance of the W(y) operator and its inverse. We note that $W(y) = W_{in}(y)$ still because $\mathbf{P}_{\Phi in}$ is independent of $y^0$.

Similarly an out $\phi$ particle can be reduced from an S matrix part. For example,

$$\begin{aligned}<\alpha\ \text{out}|W^{-1}(y)\phi(X(y))W(y)\,|\beta\ \text{in}>&=<\alpha-p'\ \text{out}|W^{-1}(y)\phi(X(y))W(y)\,|\beta-p'\ \text{in}> \\ &- i<\alpha-p'\ \text{out}|\int d^3y'\ [W^{-1}(y')\phi_{in}(X(y'))W(y')W^{-1}(y)\phi(X(y))W(y) - \\ &- W^{-1}(y)\phi(X(y))W(y)W^{-1}(y')\phi_{out}(X(y'))W(y')]\,|\beta\ \text{in}>\overleftrightarrow{\partial_0} f_{p'}^{*}(y')\end{aligned} \tag{5.77}$$

which becomes

$$\begin{aligned}<\alpha\ \text{out}|W^{-1}(y)\phi(X(y))W(y)\,|\beta\ \text{in}> &= <\alpha-p'\ \text{out}|\varphi(y)\,|\beta-p'\ \text{in}> + \\ &+ iZ^{-1/2}\int d^4y'\ <\alpha-p'\ \text{out}|T(\varphi(y')\varphi(y))\,|\beta\ \text{in}>\ (\overleftarrow{\Box}_{y'} + m^2)\ f_{p'}^{*}(y')\end{aligned} \tag{5.78}$$

where the time ordered product is defined with respect to ordering with respect to $y^0$ and where

$$\varphi(y) = W^{-1}(y)\phi(X(y))W(y) \tag{5.79}$$

These results directly generalize to multi-particle in and out states:

$$<p_1, p_2, \ldots p_n\ \text{out}|\ q_1, q_2, \ldots q_m\ \text{in}> = \ldots <0|T(\varphi(y'_1)\ \ldots\ \varphi(y'_n)\varphi(y_1)\ \ldots\ \varphi(y_m))\,|0> \ldots \tag{5.80}$$

thus reducing the development of the perturbation theory of the S matrix to the evaluation of time ordered products such as

$$<0|T(\varphi(y_1)\ \ldots\ \varphi(y_n))\,|0> \tag{5.81}$$

# 6. Perturbation Theory II

## The U Matrix

The U matrix for a two-tier theory is developed in a way similar to conventional field theory starting from the defining relations:

$$\phi(X(y)) = U^{-1}\phi_{in}(X(y))U \tag{6.1}$$

$$\pi_\phi(X(y)) = U^{-1}\pi_{\phi in}(X(y))U \tag{6.2}$$

From eq. 5.29 we define the free field hamiltonian

$$H_{F0in}(\phi_{in}, \pi_{\phi in}) = \int d^3X\, \mathscr{H}_{F0}(\phi_{in}, \pi_{\phi in}) \tag{6.3}$$

Noting $X^0 = y^0$ in the Y Coulomb gauge we find

$$\partial\phi_{in}/\partial y^0 = i[H_{F0in}, \phi_{in}(X)] \tag{6.4}$$

$$\partial\pi_{\phi in}/\partial y^0 = i[H_{F0in}, \pi_{\phi in}(X)] \tag{6.5}$$

For the entire hamiltonian (eq. 5.28) we have

$$\partial\phi/\partial y^0 = i[H_F, \phi(X)] \tag{6.6}$$

$$\partial\pi_\phi/\partial y^0 = i[H_F, \pi_\phi(X)] \tag{6.7}$$

with

$$H_F(\phi, \pi_\phi) = :\int d^3X\, \mathscr{H}_F(\phi, \pi_\phi): \tag{6.8}$$

(Note the *entire* interaction term is normal ordered since $d^3X$ is a q-number. Combining the above equations in the standard way yields a familiar differential equation for the U matrix:

$$i\partial U(y^0)/\partial y^0 = (H_{Fint} + E_0(t))U(y^0) \tag{6.9}$$

where $E_0(t)$ is a c-number function of $y^0$ that we can set equal to 0 (as it would be cancelled later in any case), and where

$$H_{Fint}(\phi_{in}, \pi_{\phi in}) = :\int d^3X\, \mathscr{H}_{Fint}(\phi_{in}, \pi_{\phi in}): \tag{6.10}$$

with $\mathscr{H}_{Fint}$ given by eq. 5.30. Solving for U gives the familiar time ordered exponential:

$$U(y^0) = T\Big(\exp[-i\int_{-\infty}^{t} dy^0\ H_{Fint}]\Big) \tag{6.11a}$$

which is a symbolic notation for:

$$U(y^0) = 1 + \sum_{n=1}^{\infty} (-i)^{-n} (n!)^{-1} \int_{-\infty}^{y^0} dy_1{}^0 \ldots \int_{-\infty}^{y^0} dy_n{}^0\, T(H_{Fint}(y_1{}^0) \ldots H_{Fint}(y_n{}^0))$$

(6.11b)

We note for later use that the hermiticity of $H_{Fint}$ is not used in the derivation of eq. 6.11. Thus eq. 6.11 would still hold if $H_{Fint}$ were not hermitean.

## Reduction of Time Ordered φ Products

In the previous chapter we reduced the calculation of the S matrix to the evaluation of time ordered products of the form

$$\tau(y_1, \ldots, y_n) = \langle 0|T(\varphi(y_1) \ldots \varphi(y_n))|0\rangle \tag{6.12}$$

where $\varphi(y)$ is specified by eq. 5.79. Expanding the terms within eq. 6.12 using eq. 5.79 we find

$$\varphi(y_1) \ldots \varphi(y_n) = W^{-1}(y_1)\phi(X(y_1))W(y_1)W^{-1}(y_2)\phi(X(y_2))W(y_2) \ldots W^{-1}(y_n)\phi(X(y_n))W(y_n) \tag{6.13}$$

which can be re-expressed as

$$W^{-1}(y_1)U^{-1}(y_1{}^0)\phi_{in}(X(y_1))U(y_1{}^0)W(y_1)W^{-1}(y_2)U^{-1}(y_2{}^0)\phi_{in}(X(y_2))U(y_2{}^0)W(y_2)\ \ldots \tag{6.14}$$

using eq. 6.1 and denoting $W_{in}(y)$ as $W(y)$. Defining

$$\mathscr{U}(y_1, y_2) = U(y_1{}^0)W(y_1)W^{-1}(y_2)U^{-1}(y_2{}^0) \tag{6.15}$$

we see eq. 6.14 can be rewritten as

$$W^{-1}(y_1)U^{-1}(y_1{}^0)\phi_{in}(X(y_1))\mathscr{U}(y_1, y_2)\phi_{in}(X(y_2))\,\mathscr{U}(y_2, y_3)\phi_{in}(X(y_3))\ \ldots\phi_{in}(X(y_n))U(y_n{}^0)W(y_n) \tag{6.16}$$

From eqs. 5.71 and 5.72

$$\mathscr{U}(y_1, y_2) = U(y_1{}^0)\exp((\mathbf{Y}(y_2) - \mathbf{Y}(y_1))\cdot\mathbf{P}_{\Phi a}/M_c^2)U^{-1}(y_2{}^0) \tag{6.17}$$

Defining

$$W(\,y_1, y_2) = \exp((\mathbf{Y}(y_2) - \mathbf{Y}(y_1))\cdot\mathbf{P}_{\Phi a}/M_c^2) \tag{6.18}$$

and looking ahead to the Wick expansion of the time ordered product of eq. 6.12 we note that the only time ordered products involving $W(y_1, y_2)$ that would appear in the expansion are

$$<0|T(\phi_{in}(X(y))W(y_1, y_2))|0> = 0 \tag{6.19a}$$

$$<0|T(Y(y)W(y_1, y_2))|0> = 0 \tag{6.19b}$$

$$<0|T(\partial Y(y)/\partial y^\mu\, W(y_1, y_2))|0> = 0 \tag{6.19c}$$

$$<0|T(\partial Y(y)/\partial y^\mu\, \phi_{in}(X(y)))|0> = 0 \tag{6.19d}$$

$$<0|T(W(y_1, y_2)W(y_3, y_4))|0> = 1 \tag{6.19e}$$

due to the factor of $\mathbf{P}_{\Phi a}$ that appears in $W(y_1, y_2)$. Also

$$<0|T(\phi_{in}(X(y))Y(y_1))|0> = 0 \tag{6.20}$$

due to the $a_{in}(p)$ and $a_{in}^{\dagger}(p)$ factors appearing in $\phi_{in}(X(y))$.

Thus the $W(y_1, y_2)$ factor in eq. 6.17 may be set to the value one with the result

$$\mathscr{U}(y_1, y_2) \equiv U(y_1^{\,0})U^{-1}(y_2^{\,0}) = U(y_1^{\,0}, y_2^{\,0}) \tag{6.21}$$

where $U(y_1^{\,0}, y_2^{\,0})$ is the conventionally defined U matrix satisfying

$$i\partial\, U(y_1^{\,0}, y_2^{\,0})/\partial y_1^{\,0} = iH_{Fint}\, U(y_1^{\,0}, y_2^{\,0}) \tag{6.22}$$

with the boundary condition

$$U(y^0, y^0) = 1 \tag{6.23}$$

This result would still be true if the $W(y_1, y_2)$ exponentials were expanded in their "power series" form.

Then, paralleling the standard approach we find an expression for the U matrix:

$$U(y_1^{\,0}, y_2^{\,0}) = T\Big(\exp[-i \int_{y_2^{\,0}}^{y1^{\,0}} dy'^0 {:}d^3X(y')\ \mathscr{H}_{Fint}(\phi_{in}(X(y')), \pi_{\phi in}(X(y'))){:}]\Big) \tag{6.24}$$

The $U(y_1^{\,0}, y_2^{\,0})$ matrix satisfies the conventional multiplication rule:

$$U(y_1^{\,0}, y_3^{\,0}) = U(y_1^{\,0}, y_2^{\,0})U(y_2^{\,0}, y_3^{\,0}) \tag{6.25}$$

The inverse of $U(y_1, y_2)$ is

$$U^{-1}(y_1^{\,0}, y_2^{\,0}) = U(y_2^{\,0}, y_1^{\,0}) \tag{6.26}$$

We now return to eq. 6.16, which can now be written in the form:

$$U^{-1}(y^0)U(y^0, y_1^{\,0})\phi_{in}(X(y_1))U(y_1^{\,0}, y_2^{\,0})\phi_{in}(X(y_2))U(y_2^{\,0}, y_3^{\,0}) \ldots \phi_{in}(X(y_n))U(y_n^{\,0}, -y^0)U(-y^0) \tag{6.27}$$

where $y^0$ is a reference time that is later than all other times, and $-y^0$ is earlier than all the other times, in the time-ordered product. As a result the time-ordered product in eq. 5.80 can be expressed in a symbolic notation as:

$$<0|U^{-1}(y^0)T(\phi_{in}(X(y_1))\phi_{in}(X(y_2)) \ldots \phi_{in}(X(y_n))U(y^0, -y^0))U(-y^0)|0> \qquad (6.28)$$

The analysis of eq. 6.28 as $y^0 \to \infty$ follows the standard path, which begins by noting

$$U(-y)|0> = \lambda_-|0> \qquad \text{when } y^0 \to \infty \qquad (6.29a)$$

$$U(y)|0> = \lambda_+|0> \qquad \text{when } y^0 \to \infty \qquad (6.29b)$$

following a standard textbook proof, which, in turn, leads to:

$$\lambda_-\lambda_+{}^* = <0|T\left(\exp[+i\int_{-\infty}^{\infty} dy'^0{:}d^3X(y')\mathscr{H}_{Fint}(\phi_{in}(X(y')), \pi_{\phi in}(X(y'))){:}]\right)|0> \qquad (6.30)$$

$$= \left[<0|T\left(\exp[-i\int_{-\infty}^{\infty} dy'^0 d^3X(y')\ \mathscr{H}_{Fint}(\phi_{in}(X(y')), \pi_{\phi in}(X(y')))]\right)|0>\right]^{-1} \qquad (6.31)$$

Thus the time ordered product of eq. 6.12, which appears in the evaluation of the S matrix element in eq. 5.80, can be symbolically written as:

$$\tau(y_1, \ldots, y_n) = \frac{<0|T(\phi_{in}(X(y_1)) \ldots \phi_{in}(X(y_n))U(\infty, -\infty))|0>}{<0|T\left(\exp[-i\int dy'^0{:}d^3X(y')\mathscr{H}_{Fint}(\phi_{in}(X(y')),\pi_{\phi in}(X(y'))){:}]\right)|0>} \qquad (6.32)$$

in the limit $y^0 \to \infty$.

## The $\int d^3X$ Integration

The integration over the X space coordinates presents the difficulty of a functional integration of a q-number that needs to be properly defined. Since

$$X^\mu(y) = y^\mu + i\, Y^\mu(y)/M_c^{\,2} \tag{3.12}$$

by definition and since, in the Y Coulomb gauge we have $X^0(y) = y^0$ due to $Y^0 = 0$, the classical Jacobian for the transformation from y to X coordinates is the absolute value:

$$J = \left| \varepsilon^{ijk} \left(\delta^{1i} + \frac{i}{M_C^{\,2}} \frac{\partial Y^1}{\partial y^i}\right)\left(\delta^{2j} + \frac{i}{M_C^{\,2}} \frac{\partial Y^2}{\partial y^j}\right)\left(\delta^{3k} + \frac{i}{M_C^{\,2}} \frac{\partial Y^3}{\partial y^k}\right)\right| \tag{6.33}$$

The Jacobian appears in a change of integration variables:

$$\int d^3X = \int d^3y\, J \tag{6.34}$$

and

$$\int d^4X = \int d^4y\, J \tag{6.35}$$

in the Y Coulomb gauge.

A change of variables for c-number coordinate transformations is well known. The situation changes when one set of coordinates are in fact q-numbers. The second quantization of the Y field requires the definition of J to be clarified since the product of fields at the same position is normally undefined. The normal ordering of the interaction hamiltonian term in eqs. 6.34 and 6.32 resolves the issue. Therefore eq. 6.33 must be considered as inserted within a normal ordered expression.

While normal ordering eliminates the infinities that would otherwise be present, J still presents a problem because it is still effectively part of the interaction term. This situation appears to be unsatisfactory in the present, scalar quantum field theory in which Y is not intended to play a direct dynamical role but rather a passive role as a coordinate. The normal $\phi$ field is supposed to be the only in, out, and interacting field.

The problem of J is resolved by eqs. 6.19c and 6.19d, which reduces the effect of the derivative terms in eq. 6.33 to zero in the Wick expansion of the time ordered product in eq. 6.32 if no Y quanta appear in or out S matrix states. Thus

$$J \equiv 1 \tag{6.36}$$

As a result the time ordered product (eq. 6.32) becomes:

$$\tau(y_1, \ldots, y_n) = \frac{<0|T(\phi_{in}(X(y_1))\ldots\phi_{in}(X(y_n))\exp[-i\int d^4y' \mathscr{H}_{Fint}(\phi_{in}(X(y')))])\ |0>}{<0|T\left(\exp\left[-i\int d^4y' \mathscr{H}_{Fint}(\phi_{in}(X(y')))\right]\right)|0>} \tag{6.37}$$

## Y In and out states

The Y fields have no interactions and are thus free fields in the model Lagrangian under consideration and in the two-tier quantum field theories that we will construct later. Therefore "in" Y quanta are the same as "out" Y quanta.

Since the Lagrangians that we consider do not have interaction terms explicitly containing Y field factors, the S matrix is "block diagonal" in the sense that if an in-state does not contain Y quanta, (or Y coherent states) then out-states will not contain Y quanta (or coherent Y states). The proof is based on the expansion of S matrix elements using Wick's theorem in products of time ordered products of pairs of in field operators. Eqs. 6.19, 6.20 and 6.36, and in particular,

$$<0|\ T(\phi_{in}(X(y_1))Y^i(y_2))\,|0> = 0 \tag{6.39}$$

and

$$<0|T(\phi_{in}(X(y_1))e^{-\mathbf{q}\cdot\mathbf{Y}^-(y)/M_c^2})\,|0> = <0|T(\phi_{in}(X(y_1))e^{+\mathbf{q}\cdot\mathbf{Y}^+(y)/M_c^2})\,|0> = 0 \tag{6.40}$$

prove S matrix elements with no incoming Y quanta or coherent states will have zero matrix elements to produce outgoing Y quanta or coherent states. In addition any non-zero S matrix element with n incoming Y quanta must have n outgoing Y quanta. For example an incoming state with 5 Y quanta and 2 $\phi$ particles can only become an outgoing state with 5 Y quanta and two or more $\phi$ particles. Therefore we have proved the general result:

**Theorem 6.I:** ***Any non-zero S matrix element has the same number of incoming Y quanta and outgoing Y quanta.***

*This theorem is true in any Two-Tier quantum field theory. In order to have a tractable theory we will require all in-states and out-states not to contain Y quanta or coherent states. All normal in-state and out-state particles will contain factors of* $:e^{\pm p\cdot Y/M_c^2}:$ *in the fourier expansions of their corresponding fields.*

## Unitarity

For many years it has been evident that modified field theories[11, 17, 28] might offer some hope of avoiding the divergences of conventional quantum field theory. Usually these theories suffer from unitarity problems: negative norms and negative probabilities. In the absence of a physically acceptable interpretation of negative probabilities, these theories have been thought to be unsatisfactory.

The two-tier type of quantum field theory *superficially* also appears to have a unitarity problem due to the non-hermitean nature of two-tier hamiltonians. The lack of hermiticity is due entirely to the appearance of $iY^{\mu}$ in the $X^{\mu}$ field coordinates. *In fact two-tier quantum field theories satisfy unitarity for physical states. Physical states are defined to consist of any number of normal two-tier particles and NO Y quanta.*

Two-tier interaction hamiltonians, such as the one in eq. 6.37, are not hermitean. For example,

$$H_{Fint} = \int d^3y' \mathscr{H}_{Fint}(\phi_{in}(y' + iY(y')/M_c^2)) \tag{6.41}$$

and

$$H_{Fint}^{\dagger} = \int d^3y' \mathscr{H}_{Fint}(\phi_{in}(y' - iY(y')/M_c^2)) \neq H_{Fint} \tag{6.42}$$

The relation between $H_{Fint}$ and its hermitean conjugate is

$$H_{Fint} = V\, H_{Fint}^{\dagger}\, V \tag{6.43}$$

where $V^2 = I$ is the metric operator defined in eqs. 5.15 – 5.18. Thus the S matrix is not unitary; the S matrix is *pseudo-unitary*:

$$S^{-1} = V\, S^{\dagger}\, V \tag{6.44}$$

and

$$VS^{\dagger}\, VS = I \tag{6.45}$$

We will now show that the S matrix is *unitary between physical states.* To prove this point, consider eq. 6.45 between physical states |i> and <f| – each consisting of a number of $\phi$ particles and no Y quanta.

---

[28] S. Blaha, Phys.Rev. **D10**, 4268 (1974); S. Blaha, Phys.Rev. **D11**, 2921 (1975); S. Blaha, Nuovo Cim. **A49**, :113 (1979); S. Blaha, "Generalization of Weyl's Unified Theory to Encompass a Non-Abelian Internal Symmetry Group" SLAC-PUB-1799, Aug 1976; S. Blaha, "Quantum Gravity and Quark Confinement" Lett. Nuovo Cim. **18**, 60 (1977); S. Blaha, "The Local Definition of Asymptotic Particle States" Nuovo Cim. **A49**, 35 (1979) and references therein.

$$\delta_{fi} = <f \mid I \mid i> = <f \mid VS^\dagger VS \mid i>$$

$$= \sum_{n,\, m,\, p} <f \mid V \mid p><p \mid S^\dagger \mid n><n \mid V \mid m><m \mid S \mid i>$$

$$= \sum_{n,\, m,\, p} <f \mid S^\dagger \mid m><m \mid S \mid i> \qquad (6.46)$$

since V has the eigenvalue 1 between states consisting of no Y quanta. Due to eqs. 6.19a – 6.19e and 6.20 since there are no incoming Y quanta there are no outgoing Y quanta. The block diagonality of S (and the diagonality of V) limits the intermediate states $|n>$ and $|m>$ to states containing $\phi$ particles and no Y quanta – although normalization factors R(**p**, z) will appear (described later) due to the presence of $:e^{\pm p\cdot Y/M_c^2}:$ factors within quantum field fourier expansions that embody Y coherent state effects. Thus

$$S_{phys}^{\dagger} S_{phys} = I \qquad (6.47)$$

and

$$S_{phys}^{\dagger} = S_{phys}^{-1} \qquad (6.48)$$

proving unitarity between physical states – states consisting of $\phi$ particles and no Y quanta that are properly normalized. A detailed example is presented starting on page 67.

*Finite Renormalization of External Legs*

In the previous section we showed the theory satisfies unitarity for states that are properly normalized. However the use of the non-unitary operator W(y) (eq. 5.6) to transform $\Phi_{in}(y)$ fields into $\phi_{in}(X(y))$ fields in the LSZ procedure in eq. 5.69, and related equations, does not preserve the norm of input and output $\phi$ particle legs. Thus a finite renormalization is needed for each external particle leg in order to have a unitary S-matrix.

We define this renormalization of external legs within the framework of a perturbation theory example in the section beginning on page 67.

## Perturbation Expansion

Perturbation theory in two-tier quantum field theory is very similar to conventional perturbation theory. The difference is in the form of the propagators, which have a high energy damping factor R(**p**, z) that eliminates infinities that normally appear at high energy in conventional quantum field theories.

In order to develop a feeling for two-tier perturbation theory we will calculate a few low order diagrams in the perturbation theory of the model scalar $\phi^4$ theory that we have been using as an example in this chapter.

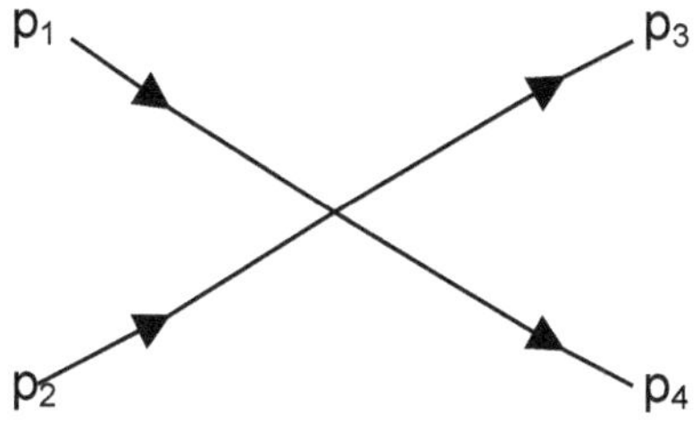


Figure 6.1. Lowest order quartic interaction diagram.

Fig. 6.1 contains the lowest order diagram for the scattering of two $\phi$ particles into a two $\phi$ particle out-state. The S matrix element for this diagram is

$$S_1 = i^4(1/4!\, i\varkappa_0) \prod_{j=1}^{4} \int d^4y_j\, d^4y\, f_{Zp_1}(y_1) f_{Zp_2}(y_2) f_{Zp_3}^{*}(y_3) f_{Zp_4}^{*}(y_4)(\Box_{y_1} + m^2)\cdot$$

$$\cdot(\Box_{y_2}+m^2)(\Box_{y_3}+m^2)(\Box_{y_4}+m^2)<0|T(\phi_{in}(X(y_1))\ldots\phi_{in}(X(y_4)):(\phi_{in}(X(y))^4:)|0> \tag{6.49}$$

with $f_{Zp}(y)$ specified by

$$f_{Zp}(y) = [(2\pi)^3 2p^0 Z_p]^{-1/2} e^{-ip\cdot y} \tag{6.49a}$$

where $Z_p$ is a normalization factor that will be specified later.

Expanding the time ordered product and realizing there are 4! ways of combining the four field factors in the interaction hamiltonian leads to:

$$S_1 = i^4(i\varkappa_0) \prod_{j=1}^{4} \int d^4y_j\, d^4y\, f_{Zp_1}(y_1) f_{Zp_2}(y_2) f_{Zp_3}^{*}(y_3) f_{Zp_4}^{*}(y_4)(\Box_{y_1} + m^2)\cdot$$

$$\cdot(\Box_{y_2}+m^2)(\Box_{y_3}+m^2)(\Box_{y_4}+m^2) i\Delta_F^{TT}(y_1-y) i\Delta_F^{TT}(y_2-y) i\Delta_F^{TT}(y_3-y) i\Delta_F^{TT}(y_4-y) \tag{6.50}$$

where

$$i\Delta_F^{TT}(y_1 - y_2) = <0|T(\phi(X(y_1)),\phi(X(y_2)))|0> \tag{6.51}$$

$$= i \int \frac{d^4p\ e^{-ip\cdot(y_1 - y_2)}\ R(\mathbf{p}, y_1 - y_2)}{(2\pi)^4\ (p^2 - m^2 + i\varepsilon)} \tag{6.52}$$

with

$$R(\mathbf{p}, y_1 - y_2) = \exp[-p^ip^j\Delta_{Tij}(y_1 - y_2)/M_c^4] \tag{6.53}$$

(summations are over space indices only in the Y Coulomb gauge) and

$$\Delta_{Tij}(z) = \int d^3k\ e^{-ik\cdot z}(\delta_{ij} - k_ik_j/\mathbf{k}^2)/[(2\pi)^3 2\omega_k] \tag{6.54}$$

From chapter 4 we have:

$$R(\mathbf{p}, y_1 - y_2) = \exp\{-p^2[A(v) + B(v)\cos^2\theta] / [4\pi^2M_c^4z^2]\} \tag{6.55}$$

with

$$z^\mu = y_1^\mu - y_2^\mu \tag{6.56}$$

$$z = |\mathbf{z}| = |\mathbf{y_1} - \mathbf{y_2}| \tag{6.57}$$

$$p = |\mathbf{p}| \tag{6.58}$$

$$v = |z^0| / z \tag{6.59}$$

$$A(v) = (1 - v^2)^{-1} + .5v\ \ln[(v - 1)/(v + 1)] \tag{6.60}$$

$$B(v) = v^2(1 - v^2)^{-1} - 1.5v\ \ln[(v - 1)/(v + 1)] \tag{6.61}$$

$$\mathbf{p}\cdot\mathbf{z} = pz\cos\theta \tag{6.62}$$

and with $|\mathbf{p}|$ denoting the length of the spatial vector $\mathbf{p}$, while $|z^0|$ is the absolute value of $z^0$.

We note

$$R(\mathbf{p}, y) = R(\mathbf{p}, -y) \tag{6.62a}$$

for later use.

Letting $y_i = w_i + y$ yields

$$S_1 = i^4(i\chi_0)(2\pi)^4\,\delta^4(p_3 + p_4 - p_1 - p_2)\,N^+(p_4)N^+(p_3)N(p_2)N(p_1) \tag{6.63}$$

where

$$N(p) = iZ_p^{-1/2}\int d^4w\; f_p(w)(\Box + m^2)\Delta_F^{TT}(w) \tag{6.64}$$

$$N^+(p) = iZ_p^{-1/2}\int d^4w\; f_p{}^*(w)(\Box + m^2)\Delta_F^{TT}(w) \tag{6.65}$$

are "normalizations" of the "external legs" – the in and out states due to the Y field cloud around each particle with $Z^{-1/2}$ a renormalization factor to be determined later. In the limit of low momentum ($p \ll M_C$):

$$N(p) = N^+(p) \rightarrow -iZ_p^{-1/2}[(2\pi)^3\, 2p^0\,]^{-1/2} \tag{6.66}$$

which the reader will note is the standard normalization factor for external scalar field legs in conventional quantum field theory modulo the $Z_p^{-1/2}$ factor. The factor $Z_p^{-1/2}$ performs the finite renormalization of external legs discussed in the preceding unitarity discussion.

*Higher Order Diagram With a Loop*

We will now consider the simplest one loop scattering diagrams in the scalar $\phi^4$ theory.

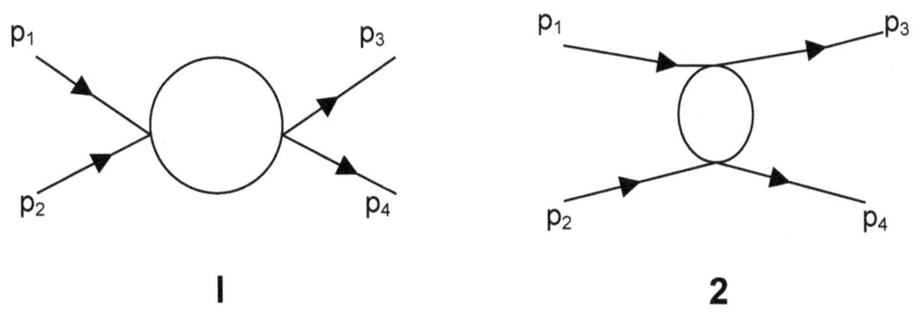


Figure 6.2. Lowest order loop scattering diagrams.

The S matrix element for these diagrams (and some other disconnected diagrams) is contained in

$$S_2 = i^4(\tfrac{1}{4!}\, i\chi_0)^2 \prod_{j=1}^{4} \int d^4y_j\, d^4y_1'\, d^4y_2'\, f_{Zp_1}(y_1) f_{Zp_2}(y_2) f_{Zp_3}{}^*(y_3) f_{Zp_4}{}^*(y_4) (\Box_{y_1} + m^2)\cdot$$
$$\cdot(\Box_{y_2} + m^2)(\Box_{y_3} + m^2)(\Box_{y_4} + m^2) <0| T(\phi_{in}(X(y_1)) \ldots \phi_{in}(X(y_4))$$
$$:(\phi_{in}(X(y_1')))^4::(\phi_{in}(X(y_2')))^4:)|0>/2! \tag{6.67}$$

together with some other disconnected diagrams.

Expanding the time ordered product and keeping only the terms corresponding to Fig. 6.2 gives:

$$S_2 = i^4(i\chi_0)^2/2 \prod_{j=1}^{4} \int d^4y_j\, d^4y_1'\, d^4y_2'\, f_{Zp_1}(y_1) f_{Zp_2}(y_2) f_{Zp_3}{}^*(y_3) f_{Zp_4}{}^*(y_4)\cdot$$
$$\cdot(\Box_{y_1} + m^2)\,(\Box_{y_2} + m^2)(\Box_{y_3} + m^2)(\Box_{y_4} + m^2)\cdot$$
$$\cdot\{i\Delta_F{}^{TT}(y_1 - y_1') i\Delta_F{}^{TT}(y_2 - y_1') i\Delta_F{}^{TT}(y_3 - y_2') i\Delta_F{}^{TT}(y_4 - y_2') +$$
$$+ i\Delta_F{}^{TT}(y_1 - y_1') i\Delta_F{}^{TT}(y_2 - y_2') i\Delta_F{}^{TT}(y_3 - y_1') i\Delta_F{}^{TT}(y_4 - y_2')\}\cdot$$
$$\cdot i\Delta_F{}^{TT}(y_1' - y_2') i\Delta_F{}^{TT}(y_1' - y_2') \tag{6.68}$$

Following a similar procedure to the previous calculation yields

$$S_2 = i^4[(i\chi_0)^2/2](2\pi)^4\delta^4(p_3 + p_4 - p_1 - p_2) N^+(p_4) N^+(p_3) N(p_2) N(p_1)\cdot$$
$$\cdot \int d^4z\, [e^{-i(p_1 + p_2)\cdot z} + e^{-i(p_1 - p_3)\cdot z}]\, [i\Delta_F{}^{TT}(z)]^2 \tag{6.69}$$

revealing a similar normalization of the external legs to that found in eq. 6.63, and a momentum conserving delta function as in eq. 6.63. The loop integrals have the form:

$$I(q) = \int d^4z\, e^{-iq\cdot z}\, [i\Delta_F{}^{TT}(z)]^2 \tag{6.70}$$

The behavior of the two-tier Feynman propagator $\Delta_F^{TT}(z)$ was studied at long and short distance in eqs. 4.21-4.24. The large distance behavior of the two-tier Feynman propagator $\Delta_F^{TT}(z)$ approaches the behavior of the conventional Feynman propagator since

$$R(\mathbf{p}, z) \to 1 \tag{6.71}$$

as $z^2 = z^\mu z_\mu$ becomes much larger than $M_c^{-2}$ ($z^2 \gg M_c^{-2}$) (eq. 6.55). Thus I(q) approaches the standard one loop expression of conventional field theory at large distance (or small momentum). Again we seee that *two-tier quantum field theory realizes a form of Correspondence Principle approaching conventional quantum field theory at large distance.*

At short distances the Gaussian factor $R(\mathbf{p}, z)$ dominates. The two-tier Feynman propagator $\Delta_F^{TT}(z)$ is radically different from the conventional Feynman propagator at very short distances (or very high momentum). The singular behavior of the conventional Feynman propagator is replaced with a well-behaved, high-energy (short distance) behavior. Near the light cone $M_c^{-2} \gg z^2 \to 0$ (or $p^2 \gg M_c^2$) we can approximate eq. 6.52 with

$$i\Delta_F^{TT}(z) \approx \int d^3p\, [N(p)]^2\, R(\mathbf{p}, z) \tag{6.72}$$

since $e^{-ip\cdot z}$ is approximately unity for small z. We assume the mass of the $\phi$ particle is negligible on this scale. Upon performing the integrations (see eq. 4.23 for the exact result) we find eq. 6.72 approaches:

$$i\Delta_F^{TT}(z) \to \pi\, M_c^4 |z^2| / 8$$

as $z^2 = z^\mu z_\mu \to 0$ from the space-like or time-like side of the light cone where $|\ |$ represents the absolute value.

Therefore I(q) is finite and well-behaved. At high energy ($q^2 \gg M_c^2$)

$$I(q) \sim q^{-8}$$

since the fourier transform of $\Delta_F^{TT}(z)$ (momentum space) is

$$\Delta_F^{TT}(p) = \int d^4z\, e^{-ip\cdot z}\, \Delta_F^{TT}(z) \sim p^{-6}$$

for large p ($p^2 \gg M_c^2$). (Compare the preceding high energy behavior of I(q) with the conventional logarithmically divergent one loop result $I(q) \sim \ln(q^2/\Lambda^2)$ with $\Lambda$ a cutoff.)

Thus two-tier quantum provides the benefits of a higher derivative theory without its drawbacks.

## Finite Renormalization of External Particle Legs & Unitarity Example

The renormalization factor $Z_p^{-1/2}$ appearing in eqs. 6.64 and 6.65 that is due to the use of the non-unitary operator W(y) (eq. 5.6) to transform $\Phi_{in}(y)$ fields into $\phi_{in}(X(y))$ fields in the LSZ procedure in eq. 5.69, and related equations, does not preserve the norm of input and output $\phi$ particle legs. $Z_p^{-1/2}$ performs a finite renormalization for each external particle leg to compensate for the effects of W(y).

The required renormalization is nicely illustrated by considering the unitarity sum in the imaginary part of the preceding example.

The transition matrix $T_{fi}$ is defined in terms of the S matrix by

$$S_{fi} = \delta_{fi} - i\,(2\pi)^4\,\delta^4(P_f - P_i)\,T^{(+)}{}_{fi}$$

The unitarity condition is

$$T^{(+)}{}_{fi} - T^{(-)}{}_{fi} = -i \sum_n\,(2\pi)^4\,\delta^4(P_n - P_i)\,T^{(-)}{}_{fn}\,T^{(+)}{}_{ni} \qquad (6.73)$$

Therefore we see that the first term on the right side of eq. 6.69 gives a transition matrix term:

$$T^{(+)}{}_{2a} = -\,i[\chi_0^{\,2}/2]N^+(p_4)N^+(p_3)N(p_2)N(p_1)\int d^4z\, e^{-iP\cdot z}[i\Delta_F^{\,TT}(z)]^2 \qquad (6.69a)$$

where $P = p_1 + p_2$. Substituting for $i\Delta_F^{\,TT}$ (using eq. 4.11) we find that the imaginary part of $T^{(+)}{}_{2a}$ is given by (Note R(**p**, z) is real.)

$$T^{(+)}{}_{2a} - T^{(-)}{}_{2a} = -i[\chi_0^{\,2}/2]N^+(p_4)N^+(p_3)N(p_2)N(p_1)\int d^4z\, e^{-iP\cdot z}\,\cdot$$
$$\cdot\,\Big[i\int d^4p\,\theta(p_0)\,\delta(p^2 - m^2)e^{-ip\cdot z}\,R(\mathbf{p}, z)/(2\pi)^3\Big]^2$$

If we express the R(**p**, z) factors in terms of their fourier transforms (see eq. 4.27):

$$R(\mathbf{p}, z) = \int d^4q\, e^{-iq\cdot z}\,R(\mathbf{p}, q)$$

Then we find

$$T^{(+)}{}_{2a} - T^{(-)}{}_{2a} = -i[\chi_0{}^2/2]N^+(p_4)N^+(p_3)N(p_2)N(p_1)\int d^4z\, e^{-iP\cdot z}\cdot$$
$$\cdot\left[i\int d^4k_1\, d^4q_1\theta(k_1{}^0)\,\delta(k_1{}^2 - m^2)e^{-ik_1\cdot z}\, e^{-iq_1\cdot z}\, R(\mathbf{k_1}, q_1)/(2\pi)^3\right]\cdot$$
$$\cdot\left[i\int d^4k_2\, d^4q_2\theta(k_2{}^0)\,\delta(k_2{}^2 - m^2)e^{-ik_2\cdot z}\, e^{-iq_2\cdot z}\, R(\mathbf{k_2}, q_2)/(2\pi)^3\right]$$

Performing the integral over z gives

$$T^{(+)}{}_{2a} - T^{(-)}{}_{2a} = +i[\chi_0{}^2/2]N^+(p_4)N^+(p_3)N(p_2)N(p_1)(2\pi)^4\cdot$$
$$\cdot\int d^4k_1d^4q_1d^4k_2d^4q_2\theta(k_1{}^0)\,\delta(k_1{}^2 - m^2)\,\theta(k_2{}^0)\,\delta(k_2{}^2 - m^2)\cdot$$
$$\cdot R(\mathbf{k_1}, q_1)R(\mathbf{k_2}, q_2)\,\delta^4(P + k_1 + q_1 + k_2 + q_2)/(2\pi)^6$$

Introducing delta functions enables us to re-express this equation as

$$T^{(+)}{}_{2a} - T^{(-)}{}_{2a} = +i[\chi_0{}^2/2]N^+(p_4)N^+(p_3)N(p_2)N(p_1)\int d^4r_1d^4r_2(2\pi)^4\delta^4(P - r_1 - r_2)\cdot$$
$$\cdot\int d^4k_1d^4q_1\,\delta^4(r_1 + k_1 + q_1)\theta(k_1{}^0)\,\delta(k_1{}^2 - m^2)\,R(\mathbf{k_1}, q_1)\cdot$$
$$\cdot\int d^4k_2d^4q_2\theta(k_2{}^0)\,\delta(k_2{}^2 - m^2)\,\delta^4(r_2 + k_2 + q_2)R(\mathbf{k_2}, q_2)/(2\pi)^6$$

which becomes

$$T^{(+)}{}_{2a} - T^{(-)}{}_{2a} = +i[\chi_0{}^2/2]N^+(p_4)N^+(p_3)N(p_2)N(p_1)\int d^4r_1d^4r_2(2\pi)^4\delta^4(P - r_1 - r_2)\cdot$$
$$\cdot\int d^4k_1\theta(k_1{}^0)\,\delta(k_1{}^2 - m^2)\,R(\mathbf{k_1}, -k_1 - r_1)\cdot$$
$$\cdot\int d^4k_2\theta(k_2{}^0)\,\delta(k_2{}^2 - m^2)R(\mathbf{k_2}, -k_2 - r_2)/(2\pi)^6$$

$R(\mathbf{k}_2, -k_2 - r_2)$ can be expressed in terms of its fourier transform $R(\mathbf{k_2}, z)$ using eq. 4.27. We can now rewrite the above expression in terms of intermediate states:

$$T^{(+)}{}_{2a} - T^{(-)}{}_{2a} = -i \int d^4r_1 d^4r_2\,(2\pi)^4\delta^4(\,P - r_1 - r_2)\cdot$$

$$\cdot\, i\chi_0 N^+(p_4)N^+(p_3)\int d^4k_1\theta(k_1{}^0)\,\delta(\,k_1{}^2 - m^2)\,\,[R(\mathbf{k_1}, -k_1 - r_1)/(2\pi)^3]\cdot$$

$$\cdot \int d^4k_2\theta(k_2{}^0)\,\delta(\,k_2{}^2 - m^2)[R(\mathbf{k_2}, -k_2 - r_2)/(2\pi)^3]i\,\chi_0 N(p_2)N(p_1)/2$$

which has the form:

$$T^{(+)}{}_{2a} - T^{(-)}{}_{2a} = -i\int d^4r_1 d^4r_2(2\pi)^4\delta^4(P - r_1 - r_2)\Big[\int d^4k_1\theta(k_1{}^0)\,\delta(k_1{}^2 - m^2)\,R(\mathbf{k_1}, -k_1 -$$
$$- r_1)/(2\pi)^3\Big]\Big[\int d^4k_2\theta(k_2{}^0)\,\delta(\,k_2{}^2 - m^2)R(\mathbf{k_2}, -k_2 - r_2)/(2\pi)^3\Big]T^{(-)}{}_{fn}T^{(+)}{}_{ni}/2!$$

where

$$T^{(-)}{}_{fn} = \chi_0 N^+(p_4)N^+(p_3)N(r_2)N(r_1)$$

$$T^{(+)}{}_{ni} = \chi_0 N^+(r_2)N^+(r_1)N(p_2)N(p_1)$$

if

$$N^+(p) = N(p) = 1 \qquad (6.74)$$

*Eq. 6.74 implies the (finite) external leg renormalization must be*

$$Z_p = -\Big[\int d^4w\ f_p(w)(\Box + m^2)\Delta_F{}^{TT}(w)\Big]^2 \qquad (6.74a)$$

*by 6.64 and 6.65. Thus all external legs must be "lopped off."*

*The result is a theory that satisfies the unitarity condition (eq. 6.73) as shown in the above detailed discussion.*

If we define

$$\mathfrak{N}(r) = \int d^4k\ \theta(k^0)\delta(\,k^2 - m^2)R(\mathbf{k}, -k - r) \qquad (6.75a)$$

$$= (2\pi)^{-4}\int d^4k\ d^4z\ \theta(k^0)\delta(\,k^2 - m^2)\ e^{-i(k\,+\,r)\cdot z}\ R(\mathbf{k}, z) \qquad (6.75b)$$

then the two-tier completeness expression becomes:

$$S_{fi} = \sum_{n} (2\pi)^{-3n}(n!)^{-1}\int\Big(\prod_{j=1}^{n} d^4r_j\,\mathfrak{N}(r_j)\Big)S_{fn}S_{ni}^{\dagger}\,\delta^4(P_n - \sum_{k=1}^{n} r_k) \qquad (6.75c)$$

This expression reflects the fact that $\phi$ particles are surrounded by a "cloud" of Y quanta. Thus we have achieved unitarity! For small momenta $r_j \ll M_c$, we find $\mathfrak{N}(r_j) \simeq \theta(r_j^0)\delta(r_j^2 - m^2)$ (eq. 6.75b with $R(k, q) \simeq 1$.) $\theta(r_j^0)\delta(r_j^2 - m^2)$ is the form seen in conventional quantum field theory. For large momenta $\mathfrak{N}(r_j)$ is very different.

## General Form of Propagators

In this chapter we have considered a scalar two-tier quantum field theory. We have seen that the two-tier Feynman propagator is well behaved near the light cone resulting in a finite $\phi^4$ theory. This finite $\phi^4$ theory approximates the results of conventional $\phi^4$ theory at low energy thus implementing a correspondence principle: *At low energy results in two-tier quantum field theory approach the corresponding results of the corresponding conventional quantum field theory.*

The observations on two-tier field theory made in this chapter generally apply to two-tier versions of Quantum Electrodynamics, ElectroWeak Theory and the Standard Model as well as two-tier Quantum Gravity:

1. At low energy ($p^2 \ll M_c^2$ or large distances $z^2 \gg M_c^{-2}$) the two-tier quantum field theory is the same as the corresponding conventional quantum field theory to good approximation. (Correspondence Principle)
2. At high energy ($p^2 \gg M_c^2$ or short distances: $z^2 \ll M_c^{-2}$) two-tier quantum field theories (of physical interest) are well-behaved and finite.
3. Two-tier quantum field theories (of physical interest) satisfy unitarity and Lorentz invariance (and in the case of quantum gravity their dynamical equations satisfy the requirements of general relativity).

The generality of these results is based on:

1. The expansion of the S matrix in time ordered products of field operators.
2. Wick's Theorem
3. The general form of all particle propagators in two-tier quantum field theories. All particle Feynman propagators have the form:

$$iG_F^{TT}{}_{\ldots}(y_1 - y_2) = \langle 0|T(\chi_{\ldots}(X(y_1)),\chi_{\ldots}(X(y_2)))|0\rangle \qquad (6.76)$$

$$= \int d^4p\, iG_{F\ldots}(p)e^{-ip\cdot(y_1 - y_2)}\, R(\mathbf{p}, y_1 - y_2) \tag{6.77}$$

where $iG_{F\ldots}(p)$ is the conventional momentum space $\chi_{\ldots}$ particle propagator, and where ... represents the relevant tensor and matrix indices. $R(\mathbf{p}, y_1 - y_2)$ introduces a damping factor in each particle propagator that eliminates divergences.

*Scalar Particle Propagator*

The two-tier propagator for the case of a free scalar particle is:

$$i\Delta_F^{TT}(y_1 - y_2) = \langle 0|\, T(\phi(X(y_1)),\phi(X(y_2)))\,|0\rangle \tag{6.51}$$

$$= i \int \frac{d^4p\, e^{-ip\cdot(y_1 - y_2)}\, R(\mathbf{p}, y_1 - y_2)}{(2\pi)^4\, (p^2 - m^2 + i\varepsilon)} \tag{6.52}$$

Since the mass m is not relevant at high energy we set $m = 0$. This enables us to obtain a more tractable expression for the propagator. After some manipulation the massless scalar propagator can be represented as:

$$i\Delta_F^{TT}(z) = -\beta[16\pi^3(AB)^{1/2}]^{-1} \int_{-\infty}^{\infty} dy_1 \int_{-\infty}^{\infty} dy_2 \cdot$$

$$\cdot\{\theta(z_0)\exp[-\beta((y_1 - z_0)^2B + (y_2 + z)^2A)/(4AB)] +$$

$$+ \theta(-z_0)\exp[-\beta((y_1 + z_0)^2B + (y_2 - z)^2A)/(4AB)]\}/(y_1^2 - y_2^2) \tag{6.78}$$

with $\beta = 4\pi^2 M_c^4 \mathbf{z}^2$. Using

$$(y_1^2 - y_2^2)^{-1} = -0.5 \int_0^{\infty} dq_1 \int_{-\infty}^{\infty} dq_2\, \theta(q_1^2 - q_2^2)\exp[iq_1y_1 - iq_2y_2] \tag{6.79}$$

we obtain the representation

$$i\Delta_F^{TT}(z^\mu) = (8\pi^2)^{-1} \int_0^{\infty} dq_1 \int_{-\infty}^{\infty} dq_2\, \theta(q_1^2 - q_2^2) \cdot$$

$$\cdot \exp\{iq_1|z_0| + iq_2z - [A'q_1^2 + B'q_2^2]/[\beta'(\mathbf{z}^2 - z_0^2)]\} \quad (6.80)$$

where $|z_0|$ is the absolute value of $z_0$, $z^2 - z_0^2 = -z^\mu z_\mu$ and

$$A = A'/(1 - v^2) \quad (6.81)$$

$$B = B'/(1 - v^2) \quad (6.82)$$

$$\beta = 4\pi^2 M_c^4 \mathbf{z}^2 = \beta' z^2 \quad (6.83)$$

with $\mathbf{z} = |\vec{\mathbf{z}}|$ – the magnitude of the spatial vector $\vec{\mathbf{z}}$, and A and B given by eqs. 6.60 – 6.61.

The representation of $i\Delta_F^{TT}$ in eq. 6.80 is particularly useful in determining its low energy ($\ll M_c$), and its high energy ($\gg M_c$) behavior. The low energy behavior is governed by the linear terms in the exponential in eq. 6.80 since $\beta'(\mathbf{z}^2 - z_0^2)$ is very large in this limit:

$$i\Delta_F^{TT}(z^\mu)_{low} \simeq (8\pi^2)^{-1} \int_0^\infty dq_1 \int_{-\infty}^\infty dq_2\, \theta(q_1^2 - q_2^2)\exp\{iq_1|z_0| + iq_2z\} \quad (6.84)$$

$$= [4\pi^2(\mathbf{z}^2 - z_0^2)]^{-1} \quad (6.85)$$

$$= i\Delta_F(z^\mu)$$

equaling the exact massless, free, spin 0 Feynman propagator of conventional quantum field theory.

In the high energy limit when $\beta'(\mathbf{z}^2 - z_0^2)$ is small since $\mathbf{z}^2 \approx z_0^2$ (i.e. near the light cone), the quadratic terms in the exponential in eq. 6.80 dominate and $A' \simeq B'$. We then find

$$i\Delta_F^{TT}(z^\mu)_{high} \simeq (8\pi^2)^{-1} \int_0^\infty dq_1 \int_{-\infty}^\infty dq_2 \theta(q_1^2 - q_2^2)\exp\{A'(q_1^2 + q_2^2)/[\beta'(\mathbf{z}^2 - z_0^2)]\} \quad (6.86)$$

$$= \pi M_c^4 |(\mathbf{z}^2 - z_0^2)|/8 \quad (6.87)$$

as in eq. 4.24. As pointed out earlier, eq. 6.87 corresponds to $k^{-6}$ behavior in momentum space:

$$i\Delta_F^{TT}(k)_{high} \backsim k^{-6} \tag{6.87a}$$

*Spin ½ Particle Propagator*

For the case of a free, spin ½ particle the propagator is:

$$iS_F^{TT}(y_1 - y_2) = <0|T(\bar{\psi}(X(y_1))\psi(X(y_2)))|0> \tag{6.88}$$

$$= i\int \frac{d^4p\, e^{-ip\cdot(y_1 - y_2)}\,(\not{p} + m)\, R(\mathbf{p}, y_1 - y_2)}{(2\pi)^4\,(p^2 - m^2 + i\varepsilon)}$$

Again setting m = 0 we find a convenient representation in the form:

$$S_F^{TT}(z^\mu) = i(8\pi^2)^{-1}\int_0^\infty dq_1 \int_{-\infty}^\infty dq_2\, \theta(q_1^{\,2} - q_2^{\,2})(\in(z_0)q_1\gamma_0 - q_2\vec{\mathbf{z}}\cdot\vec{\gamma}/z)\cdot$$

$$\cdot\exp\{iq_1|z_0| + iq_2 z - [A'q_1^{\,2} + B'q_2^{\,2}]/[\beta'(\mathbf{z}^2 - z_0^{\,2})]\} \tag{6.89}$$

using the same symbols and notation as eq. 6.80, and with $\in(z_0) = +1$ if $z_0 \geq 0$ and $-1$ otherwise.

The representation of $S_F^{TT}$ in eq. 6.89 is useful in determining its low energy ($\ll M_c$), and high energy ($\gg M_c$) behavior. The low energy behavior is governed by the linear terms in the exponential in eq. 6.89 since $\beta'(\mathbf{z}^2 - z_0^{\,2})$ is large in this limit:

$$S_F^{TT}(z^\mu)_{low} \simeq (8\pi^2)^{-1}\int_0^\infty dq_1 \int_{-\infty}^\infty dq_2\, \theta(q_1^{\,2} - q_2^{\,2})(\in(z_0)q_1\gamma_0 - q_2\vec{\mathbf{z}}\cdot\vec{\gamma}/z)\cdot$$

$$\cdot\exp\{iq_1|z_0| + iq_2 z\} \tag{6.90}$$

$$= \not{z}[2\pi^2(\mathbf{z}^2 - z_0^{\,2})^2]^{-1} \tag{6.91}$$

$$= S_F(z^\mu)$$

equaling the exact massless, spin ½ Feynman propagator of conventional quantum field theory. If we had not set m = 0 initially, we would have obtained the usual massive, spin ½ Feynman propagator.

In the high energy limit when $\beta'(\mathbf{z}^2 - z_0^2)$ is small since $\mathbf{z}^2 \approx z_0^2$ (i.e. near the light cone), the quadratic terms in the exponential in eq. 6.89 dominate and $A' \simeq B'$. We then find

$$S_F^{TT}(z^\mu)_{high} \simeq (8\pi^2)^{-1} \int_0^\infty dq_1 \int_{-\infty}^\infty dq_2 \theta(q_1^2 - q_2^2)(\epsilon(z_0) q_1 \gamma_0 - q_2 \vec{\mathbf{z}} \cdot \vec{\gamma} / z) \cdot$$

$$\cdot \exp\{A'(q_1^2 + q_2^2) / [\beta'(\mathbf{z}^2 - z_0^2)]\} \tag{6.92}$$

$$= i(8\pi^2)^{-1}\{\mathbf{z}^{-1}(\mathbf{z}^2 - z_0^2)^{3/2} 2^{3/2} \pi^{7/2} M_c^6 z_0 \gamma_0 - $$
$$- 4i(\mathbf{z}^2 - z_0^2)^2 \pi^5 M_c^8 \vec{\mathbf{z}} \cdot \vec{\gamma})\} \tag{6.93}$$

The leading momentum dependence of the fourier transform of $S_F^{TT}(z^\mu)_{high}$ is

$$S_F^{TT}(p)_{high} \backsim M_c^6 p^{-7} \gamma_0 \tag{6.94}$$

*Massless Spin 1 Particle Propagator*

The two-tier Feynman propagator for the case of a free, massless, spin 1, gauge field particle (coupled to a conserved current) such as a photon is:

$$iD_F^{TT}(z)_{\mu\nu} = -i \int \frac{d^4p\, e^{-ip\cdot z} g_{\mu\nu} R(\mathbf{p}, y_1 - y_2)}{(2\pi)^4 (p^2 + i\varepsilon)} \tag{6.95}$$

The form of eq. 6.95 is the same as the scalar particle propagator multiplied by $-g_{\mu\nu}$. As a result we have the representation:

$$iD_F^{TT}(z)_{\mu\nu} = -(8\pi^2)^{-1} \int_0^\infty dq_1 \int_{-\infty}^\infty dq_2\, \theta(q_1^2 - q_2^2)\, g_{\mu\nu} \cdot$$

$$\cdot\ \exp\{iq_1|z_0| + iq_2z - [A'q_1^2 + B'q_2^2]/[\beta'(\mathbf{z}^2 - z_0^2)]\} \quad (6.96)$$

As before in the scalar particle case, the low energy behavior is governed by the linear terms in the exponential in eq. 6.96 since $\beta'(\mathbf{z}^2 - z_0^2)$ is very large in this limit:

$$iD_F^{TT}(z)_{\mu\nu\text{low}} \simeq -g_{\mu\nu}(8\pi^2)^{-1}\int_0^\infty dq_1 \int_{-\infty}^\infty dq_2\, \theta(q_1^2 - q_2^2)\exp\{iq_1|z_0| + iq_2z\} \quad (6.97)$$

$$= -g_{\mu\nu}[4\pi^2(\mathbf{z}^2 - z_0^2)]^{-1} \quad (6.98)$$

$$= -ig_{\mu\nu}\Delta_F(z)$$

equaling the exact free, massless, spin 1 Feynman gauge field propagator of conventional quantum field theory.

In the high energy limit when $\beta'(\mathbf{z}^2 - z_0^2)$ is small since $\mathbf{z}^2 \approx z_0^2$ (i.e. near the light cone), the quadratic terms in the exponential in eq. 6.96 dominate, and $A' \simeq B'$. We then find

$$iD_F^{TT}(z)_{\mu\nu\text{high}} \simeq -(8\pi^2)^{-1}\int_0^\infty dq_1 \int_{-\infty}^\infty dq_2\theta(q_1^2 - q_2^2)g_{\mu\nu}\exp\{\, A'(q_1^2 + q_2^2)/[\beta'(\mathbf{z}^2 - z_0^2)]\} \quad (6.99)$$

$$= -g_{\mu\nu}\pi\, M_c^4\, |(\mathbf{z}^2 - z_0^2)|/8 \quad (6.100)$$

Eq. 6.100 corresponds to $k^{-6}$ behavior in momentum space:

$$iD_F^{TT}(k)_{\mu\nu\text{high}} \backsim g_{\mu\nu}\, M_c^4 k^{-6} \quad (6.101)$$

*Spin 2 Particle Propagator*

The two-tier propagator for the case of a free, massless, spin 2 particle such as a graviton is:

$$i\Delta_{F2}^{TT}(z)_{\mu\nu\rho\sigma} = i\int \frac{d^4p\, e^{-ip\cdot z}\, b_{\mu\nu\rho\sigma}(p)R(\mathbf{p}, y_1 - y_2)}{(2\pi)^4\,(p^2 + i\varepsilon)} \quad (6.102)$$

in an appropriate gauge where $b_{\mu\nu\rho\sigma}(p)$ is a tensor that is independent of the coordinates. We can express eq. 6.102 in the form:

$$i\Delta_{F2}{}^{TT}(z)_{\mu\nu\rho\sigma} = (8\pi^2)^{-1} \int_0^\infty dq_1 \int_{-\infty}^\infty dq_2\, \theta(q_1{}^2 - q_2{}^2)\, \tilde{b}(z_0, z, q_1, q_2)_{\mu\nu\rho\sigma} \cdot$$

$$\cdot \exp\{ iq_1|z_0| + iq_2 z - [A'q_1{}^2 + B'q_2{}^2]/[\beta'(\mathbf{z}^2 - z_0{}^2)]\} \quad (6.103)$$

where $\tilde{b}(z_0, z, q_1, q_2)_{\mu\nu\rho\sigma}$ is a tensor generated from the $b_{\mu\nu\rho\sigma}(p)$ tensor.

As before in the scalar particle case, the low energy behavior is governed by the linear terms in the exponential in eq. 6.103 since $\beta'(\mathbf{z}^2 - z_0{}^2)$ is very large in this limit and we find that the covariant piece[29] behaves like:

$$i\Delta_{F2}{}^{TT}(z)_{\mu\nu\rho\sigma\mathrm{lowCov}} \simeq \tilde{\tilde{b}}_{\mu\nu\rho\sigma}(8\pi^2)^{-1} \int_0^\infty dq_1 \int_{-\infty}^\infty dq_2\, \theta(q_1{}^2 - q_2{}^2)\exp\{ iq_1|z_0| + iq_2 z\} \quad (6.104)$$

$$= \tilde{\tilde{b}}_{\mu\nu\rho\sigma}[4\pi^2(\mathbf{z}^2 - z_0{}^2)]^{-1} \quad (6.105)$$

$$= i\Delta_F(z^\mu)\, \tilde{\tilde{b}}_{\mu\nu\rho\sigma}$$

where

$$\tilde{\tilde{b}}_{\mu\nu\rho\sigma} = \tfrac{1}{2}\, [\eta_{\mu\rho}\eta_{\nu\sigma} + \eta_{\mu\sigma}\eta_{\nu\rho} - \eta_{\mu\nu}\eta_{\rho\sigma}] \quad (6.106)$$

so that the expression in eq. 6.105 equals the corresponding covariant piece of the exact free, massless, spin 2 Feynman propagator of conventional quantum field theory.

In the high energy limit when $\beta'(\mathbf{z}^2 - z_0{}^2)$ is small since $\mathbf{z}^2 \approx z_0{}^2$ (i.e. near the light cone), the quadratic terms in the exponential in eq. 6.103 dominate, and $A' \simeq B'$. We then find

$$i\Delta_{F2}{}^{TT}(z)_{\mu\nu\rho\sigma\mathrm{high}} \simeq (8\pi^2)^{-1} \int_0^\infty dq_1 \int_{-\infty}^\infty dq_2 \theta(q_1{}^2 - q_2{}^2)\;\; \tilde{b}(z_0, z, q_1, q_2)_{\mu\nu\rho\sigma} \cdot$$

$$\cdot \exp\{A'(q_1{}^2 + q_2{}^2)/[\beta'(\mathbf{z}^2 - z_0{}^2)]\} \quad (6.107)$$

and the covariant piece behaves like

---

[29] S. Weinberg, Phys. Rev. **135**, B1049 (1964); Phys. Rev. **138**, B988 (1965).

$$i\Delta_{F2}{}^{TT}(z)_{\mu\nu\rho\sigma\text{highCov}} \simeq \tilde{\tilde{b}}_{\mu\nu\rho\sigma}\pi M_c{}^4 \,|\,(z^2 - z_0{}^2)\,|\,/8 \qquad (6.108)$$

The coordinate space behavior of eq. 6.108 corresponds to $k^{-6}$ behavior in momentum space:

$$i\Delta_{F2}{}^{TT}(k)_{\mu\nu\rho\sigma\text{highCov}} \backsim \tilde{\tilde{b}}_{\mu\nu\rho\sigma}\, k^{-6} \qquad (6.109)$$

The high-energy behavior of the spin 2 propagator in momentum space results in a two-tier theory of quantum gravity that has no high-energy divergences and is thus finite. See chapter 9 for a detailed discussion.

# 7. Two-Tier Quantum Electrodynamics

## Formulation

There have been numerous attempts to develop a finite theory of Quantum Electrodynamics (QED). Among the noteworthy attempts are the Lee-Wick[30] formulation of QED and the Johnson-Baker-Willey model.[31] The unification of the electromagnetic interaction with the weak interaction in the Electroweak Theory, and the proof that it is renormalizable, has switched the focus of interest away from QED. However the extremely precise experimental tests of QED, which are among the most accurate measurements made by science, and the impressive agreement[32] with the theoretical predictions of QED, make QED of interest in its own right.

In this chapter we will describe the formulation of two-tier QED. We will see that it is finite, and yet it is in complete agreement with the highly accurate calculations of QED if $M_c$ is sufficiently large – such as of the order of the Planck mass. We will also see a modification of the Coulomb potential in the two-tier model that makes possible the existence of (unstable) bound states of particles with the same charge such as a two electron bound state. The two-tier QED Coulomb potential is linear at ultra-short distances and zero at $r = 0$.

## A New Quantum Electrodynamics with Non-Commuting Coordinates

Two-tier QED is formulated in a way that is similar to conventional QED and captures the excellent results of QED while making the theory finite. We will consider the case of QED for electrons. The results apply directly to any charged spin ½ field and with a few changes to charged particles of other spin. The two-tier QED Lagrangian that we will investigate is:

---

[30] T. D. Lee and G. C. Wick, Phys. Rev. **D2**, 1033 (1970); T. D. Lee and G. C. Wick, Nucl. Phys. **B9**, 209 (1969) and references therein.

[31] S. Blaha, "An Approximate Calculation of the Eigenvalue Function in Massless Quantum Electrodynamics", Phys.Rev. **D9**, 2246 (1974) and references therein.

[32] T. Kinoshita, "The Fine Structure Constant", Cornell University preprint CLNS 96/1406 (1996); V. W. Hughes and T. Kinoshita, Rev. Mod. Phys. **71**, S133 (1999).

$$\mathscr{L} = \mathrm{J}\mathscr{L}_{\mathrm{F}} + \mathscr{L}_{\mathrm{C}}(\mathrm{X}^{\mu}(\mathrm{y}), \partial \mathrm{X}^{\mu}(\mathrm{y})/\partial \mathrm{y}^{\nu}, \mathrm{y}) \tag{7.1}$$

with J the Jacobian and

$$\mathscr{L}_{\mathrm{F}} = \bar{\psi}(\mathrm{X}(\mathrm{y}))((\mathrm{i}\nabla\!\!\!/_{\mathrm{X}} - \mathrm{e}_0 A\!\!\!/(\mathrm{X}(\mathrm{y})) - \mathrm{m}_0)\psi(\mathrm{X}(\mathrm{y})) - \tfrac{1}{4}\, \mathrm{F}^{\mu\nu}(\mathrm{X}(\mathrm{y}))\mathrm{F}_{\mu\nu}(\mathrm{X}(\mathrm{y})) \tag{7.2}$$

with

$$\mathrm{F}_{\mu\nu}(\mathrm{X}(\mathrm{y})) = \partial \mathrm{A}_{\mu}(\mathrm{X}(\mathrm{y}))/\partial \mathrm{X}^{\nu} - \partial \mathrm{A}_{\nu}(\mathrm{X}(\mathrm{y}))\,/\partial \mathrm{X}^{\mu} \tag{7.3}$$

and

$$\mathscr{L}_{\mathrm{C}}(\mathrm{X}^{\mu}(\mathrm{y}), \partial \mathrm{X}^{\mu}(\mathrm{y})/\partial \mathrm{y}^{\nu}, \mathrm{y}) = -\tfrac{1}{4}\, \mathrm{F}_{\mathrm{Y}}{}^{\mu\nu}\mathrm{F}_{\mathrm{Y}\mu\nu} \tag{7.4}$$

$$\mathrm{F}_{\mathrm{Y}\mu\nu} = \partial \mathrm{Y}_{\mu}/\partial \mathrm{y}^{\nu} - \partial \mathrm{Y}_{\nu}\,/\partial \mathrm{y}^{\mu} \tag{7.5}$$

We note the Lagrangian $\mathscr{L}_{\mathrm{F}}$ has the form of the conventional electromagnetic Lagrangian[33] except for the functional dependence on X(y).

Since the Lagrangian in eq. 7.2 is separable we will follow the same procedure as we did for the scalar field theory in the development of eqs. A.74 – A.112. Thus we obtain the hamiltonian (as in eq. A.112 for the scalar case):

$$\mathrm{H}_{\mathrm{F}} = :\!\int \mathrm{d}^3\mathrm{X}\, (\mathscr{H}_{\mathrm{F0}} + \mathscr{H}_{\mathrm{Fint}})\!: \tag{7.6}$$

where

$$\mathscr{H}_{\mathrm{F0}} = \bar{\psi}(\mathrm{X}(\mathrm{y}))(\mathrm{i}\nabla\!\!\!/_{\mathrm{X}} - \mathrm{m}_0)\psi(\mathrm{X}(\mathrm{y})) + \tfrac{1}{2}(\mathbf{E}^2 + \mathbf{B}^2) \tag{7.7}$$

$$\mathscr{H}_{\mathrm{Fint}} = \mathrm{e}_0\bar{\psi}(\mathrm{X}(\mathrm{y}))A\!\!\!/(\mathrm{X}(\mathrm{y}))\psi(\mathrm{X}(\mathrm{y})) \tag{7.8}$$

with **E** being the electric field and **B** being the magnetic field:

$$\mathrm{E}^{\mathrm{i}} = -\partial \mathrm{A}^{\mathrm{i}}/\partial \mathrm{y}^0$$

$$\mathrm{B}^{\mathrm{i}} = \varepsilon^{\mathrm{ijk}}\, \partial \mathrm{A}_{\mathrm{j}}/\partial \mathrm{y}^{\mathrm{k}}$$

The field equations are:

---

33 We follow the conventions of Kaku (1993), and Bjorken (1995). It will be evident that the proof that two-tier QED is finite will not require detailed knowledge of specific conventions.

$$(i\not{\nabla}_X - e_0\not{A}(X(y)) - m_0)\psi(X(y)) = 0 \tag{7.9}$$

and

$$\partial(F^{\mu\nu}(X(y)))/\partial X^\nu = e_0\bar{\psi}(X(y))\gamma^\mu\psi(X(y)) \tag{7.10}$$

## The Y Field

The $X^\mu$ coordinate field is related to the $Y^\mu$ field via:

$$X_\mu(y) = y_\mu + i\,Y_\mu(y)/M_c^2 \tag{3.12}$$

The Y field has the free Lagrangian eq. 7.4 (based on eqs. 3.10 – 3.14), and the hamiltonian

$$\mathscr{H}_C = \frac{1}{2}(E_Y^2 + B_Y^2) \tag{7.11}$$

where

$$E_Y^i = -\partial Y^i/\partial y^0 \tag{7.12}$$

$$B_Y^i = \varepsilon^{ijk}\,\partial Y_j/\partial y^k \tag{7.13}$$

The Y field equations are

$$\partial F_Y^{\mu\nu}(y)/\partial y^\nu = 0 \tag{7.14}$$

The quantization of the Y field in the Coulomb gauge is described in chapter 3. We will use the Y Coulomb gauge throughout our discussions of various two-tier quantum field theories.

## Quantization of the Free Dirac Field

The quantization procedure is formally identical to that of conventional QED. The standard equal time anti-commutation relations for the spin ½ field are:

$$\{\psi_\alpha(X), \psi_\beta(X')\} = \{\pi_{\psi\alpha}(X), \pi_{\psi\beta}(X')\} = 0 \tag{7.15}$$

$$\{\pi_{\psi\alpha}(X), \psi_\beta(X')\} = i\,\delta_{\alpha\beta}\,\delta^3(\mathbf{X} - \mathbf{X}') \tag{7.16}$$

where $\alpha$ and $\beta$ are the spinor indices and where

$$\pi_{\psi a}(X) = i\,\psi_a^{\dagger}(X) \tag{7.17}$$

The spin ½ field can be expanded in a fourier series:

$$\psi(X(y)) = \sum_{\pm s}\int d^3p\; N^d_{\ m}(p)\; [b(p,s)u(p,s)\; :e^{-ip\cdot(y\,+\,iY/M_c^{\,2})}: + \\ + d^{\dagger}(p,s)v(p,s)\; :e^{ip\cdot(y\,+\,iY/M_c^{\,2})}:] \tag{7.18}$$

$$\psi^{\dagger}(X(y)) = \sum_{\pm s}\int d^3p\; N^d_{\ m}(p)\; [b^{\dagger}(p,s)\bar{u}(p,s)\gamma^0\; :e^{+ip\cdot(y\,+\,iY/M_c^{\,2})}: + \\ + d(p,s)\bar{v}(p,s)\gamma^0\; :e^{-ip\cdot(y\,+\,iY/M_c^{\,2})}:] \tag{7.19}$$

where

$$N^d_{\ m}(p) = [m/((2\pi)^3E_p)]^{1/2} \tag{7.20}$$

and

$$E_p = (\mathbf{p}^2 + m^2)^{1/2} \tag{7.21}$$

The commutation relations of the Fourier coefficient operators are:

$$\{b(p,s),\, b^{\dagger}(p',s')\} = \delta_{ss'}\delta^3(\mathbf{p} - \mathbf{p}') \tag{7.22}$$

$$\{d(p,s),\, d^{\dagger}(p',s')\} = \delta_{ss'}\delta^3(\mathbf{p} - \mathbf{p}') \tag{7.23}$$

$$\{b(p,s),\, b(p',s')\} = \{d(p,s),\, d(p',s')\} = 0 \tag{7.24}$$

$$\{b^{\dagger}(p,s),\, b^{\dagger}(p',s')\} = \{d^{\dagger}(p,s),\, d^{\dagger}(p',s')\} = 0 \tag{7.25}$$

$$\{b(p,s),\, d^{\dagger}(p',s')\} = \{d(p,s),\, b^{\dagger}(p',s')\} = 0 \tag{7.26}$$

$$\{b^{\dagger}(p,s),\, d^{\dagger}(p',s')\} = \{d(p,s),\, b(p',s')\} = 0 \tag{7.27}$$

The spinors u(p,s) and v(p,s) are defined in the conventional way (as in Kaku (1993), and in Bjorken (1965)).

## Quantization of the Electromagnetic Field

The gauge invariance of two-tier QED can be seen by examining the field equation

$$(i\nabla\!\!\!/_X - e_0 A\!\!\!/(X(y)) - m_0)\psi(X(y)) = 0 \qquad (7.9)$$

and considering the effect of a gauge gauge transformation:

$$A^\mu(X(y)) \rightarrow A^\mu(X(y)) - \partial\Lambda(X(y))/\partial X_\mu \qquad (7.28)$$

The field equation eq. 7.10 remains unchanged and the change in eq. 7.9 can be accommodated by a change of phase of the Dirac field:

$$\psi(X(y)) \rightarrow \exp(ie_0\Lambda(X(y))\ \psi(X(y)) \qquad (7.29)$$

The only novelty is that $\Lambda(X(y))$ in general becomes a complex q-number quantity at extremely short distances since X is a complex q-number.

The gauge invariance of the Lagrangian eq. 7.2 allows us to choose a convenient gauge. It appears that the most convenient gauge is the Coulomb gauge[34]:

$$\partial A^i/\partial X^i = 0 \qquad (7.30)$$

where the sum is over spatial components labeled with i. We also set

$$A^0 = 0 \qquad (7.31)$$

in the absence of an external source.
A conventional treatment leads to the equal time commutation relations:

$$[A^\mu(X(y)), A^\nu(X(y'))] = [\pi_A{}^\mu(X(y)), \pi_A{}^\nu(X(y'))] = 0 \qquad (7.32)$$

$$[\pi_A{}^j(X(y)), A_k(X(y'))] = -i\,\delta^{tr}{}_{jk}(\mathbf{X}(y) - \mathbf{X}(y')) \qquad (7.33)$$

(note the locations of the j component label in eq. 7.33 introduces a minus sign) where

---

34 It is also possible to quantize in two-tier QED using an indefinite metric that preserves manifest Lorentz covariance as was done by Gupta and Bleuler for conventional QED. See Heitler (1954) or Bogoliubov (1959).

$$\pi_A{}^k = \partial \mathscr{L}_F / \partial \dot{A}_k \tag{7.34}$$

$$\pi_A{}^0 = 0 \tag{7.35}$$

$$\delta^{tr}{}_{jk}(X(y) - X(y')) = \int d^3k\, e^{i\,\mathbf{k}\cdot(\mathbf{X}(y) - \mathbf{X}(y'))}\, (\delta_{jk} - k_j k_k / \mathbf{k}^2)/(2\pi)^3 \tag{7.36}$$

$$\dot{A}_k = \partial A_k / \partial X^0 \tag{7.37}$$

The Coulomb gauge reveals the two transverse degrees of freedom that are present in the vector potential. The Fourier expansion of the vector potential is:

$$A^i(X(y)) = \int d^3k\, N_0(k) \sum_{\lambda=1}^{2} \varepsilon^i(k, \lambda)[a(k,\lambda)\, e^{-ik\cdot X(y)} + a^\dagger(k,\lambda)\, e^{ik\cdot X(y)}] \tag{7.38}$$

where

$$N_0(k) = [(2\pi)^3 2\omega_k]^{-1/2} \tag{7.39}$$

(m = 0) and

$$\omega_k = (\mathbf{k}^2)^{1/2} = k^0 \tag{7.40}$$

with $\in(k, \lambda)$ being the polarization unit vectors for $\lambda = 1, 2$.

The commutation relations of the Fourier coefficient operators are:

$$[a(k,\lambda), a^\dagger(k',\lambda')] = \delta_{\lambda\lambda'} \delta^3(\mathbf{k} - \mathbf{k}') \tag{7.41}$$

$$[a^\dagger(k,\lambda), a^\dagger(k',\lambda')] = [a(k,\lambda), a(k',\lambda')] = 0 \tag{7.42}$$

and the polarization vectors satisfy

$$\sum_{\lambda=1}^{2} \varepsilon_i(k, \lambda) \varepsilon_j(k, \lambda) = (\delta_{ij} - k_i k_j / \mathbf{k}^2) \tag{7.43}$$

## Particle propagators

The electron and photon Feynman propagators differ from the conventional QED propagators by having the Gaussian factor $R(\mathbf{p}, z)$ in their fourier expansions:

$$iS_F^{TT}(y_1 - y_2) = <0|T(\bar{\psi}(X(y_1))\psi(X(y_2)))|0> \tag{7.44}$$

where the time ordering is with respect to $y_1^0$ and $y_2^0$. Expanding the free fields leads to the fourier representation:

$$iS_F^{TT}(y_1 - y_2) = i \int \frac{d^4p\, e^{-ip\cdot(y_1 - y_2)} (\not{p} + m)\, R(\mathbf{p}, y_1 - y_2)}{(2\pi)^4 (p^2 - m^2 + i\varepsilon)} \tag{7.45}$$

with the Gaussian factor $R(\mathbf{p}, z)$ specified in eq. 6.53. The photon propagator is

$$iD_F^{trTT}(y_1 - y_2)_{\mu\nu} = <0|T(A_\mu(X(y_1))A_\nu(X(y_2)))|0> \tag{7.46}$$

$$= -ig_{\mu\nu} \int \frac{d^4k\, e^{-ik\cdot(y_1 - y_2)} R(\mathbf{k}, y_1 - y_2)}{(2\pi)^4 (k^2 + i\varepsilon)} \tag{7.47}$$

plus gauge terms and minus the Coulomb term. The presence of the Gaussian factor $R(\mathbf{p}, z)$ results in a theory of QED that has no divergences and thus is finite. See eqns. 6.94 and 6.101 for their large momentum behavior.

## Coulomb Interaction

The Coulomb interaction in two-tier QED is different at short distances from the conventional Coulomb interaction. The Coulomb potential between singly charged (same sign) particles in two-tier QED is:

$$V_{TT}(y_1 - y_2) = e^2 \int \frac{d^4p\, e^{-ip\cdot(y_1 - y_2)} R(\mathbf{p}, y_1 - y_2)}{(2\pi)^4\, \mathbf{k}^2} \tag{7.48}$$

$$= \alpha\, \Phi(M_c^2 \pi r^2)\, \delta(y_1^0 - y_2^0)/r \tag{7.49}$$

where $a$ is the fine structure constant, $\Phi(z)$ is the error function, $M_c$ is the mass setting the scale of the short distance behavior, and $r = \surd(\mathbf{y}_1 - \mathbf{y}_2)^2$ is the radial distance. At small distances ($\pi r^2 \ll M_c^{-2}$) the two-tier potential becomes linear in r:

$$V_{TT} \to 2ar\surd\pi\, M_c^{\,2}\delta(y_1^{\,0} - y_2^{\,0}) \tag{7.50}$$

and at large distances ($\pi r^2 \gg M_c^{-2}$) the two-tier potential approaches the conventional Coulomb potential:

$$V_{TT} \to V_{Coul} = a\,\delta(y_1^{\,0} - y_2^{\,0})/r \tag{7.51}$$

using the error function normalization $\Phi(\infty) = 1$. The modified Coulomb potential $V_{TT}$ of eq. 7.49 (modulo the delta function in time) is plotted in Fig. 7.5 using $M_c = 200$ GeV/$c^2$ and Fig 7.6 using $M_c$ = Planck mass. At large distances the Coulomb potential (which has been verified experimentally with great precision) can be approximated arbitrarily closely by two-tier QED by simply letting $M_c$ become larger. Conceivably $M_c$ can be as large as the Planck mass ($1.221 \times 10^{19}$ GeV/$c^2$) or even larger. Thus conventional QED is the "large" distance limit of two-tier QED.

The short distance behavior of the two-tier Coulomb potential opens the possibility of quasi-bound states of particles of the same sign such as a two electron bound state. The normally repulsive potential has a linear behavior near $r = 0$ and a potential barrier before becoming like the conventional Coulomb potential at larger distances. A pair of electrons, if localized within the linear region of the potential, would be bound but the "bound state" would quickly decay via electron tunneling through the barrier. States of this type conceivably might have existed in the first instants after the Big Bang and influenced the earliest evolution of the universe. Creating these dilepton states does not appear to be feasible if $M_c$ is extremely large.

## Asymptotic States

The development of perturbation theory in chapter 6 applies to the two-tier theory of QED with only superficial changes.

First we note that the form of the photon propagator has exactly the form of eq. 7.47 since terms proportional to $k_\mu$ or $k_\nu$ that would appear in the evaluation of eq. 7.46 do not contribute due to current conservation. In addition the instantaneous Coulomb interaction cancels the remaining Coulomb-like term appearing in the evaluation of eq. 7.46.

Thus we are left with the electron propagator (eq. 7.45) and the effective photon propagator (eq. 7.47). The formalism and role of the Y field is the same as in the scalar two-tier quantum field theory considered earlier.

*In-states and Out-states*

The dependence of the Lagrangian, and the particle fields in particular, on $X^\mu$ rather than directly on the coordinates $y^\mu$ leads to a "fuzziness" of the definition of asymptotic particle states that we have chosen to resolve with the construction of an asymptotic free field for each particle species of the "normal" sort. In the scalar case we defined an auxiliary field $\Phi_{in}(y)$ using the creation and annihilation operators of the free scalar field $\phi_{in}(X(y))$. In actuality $\Phi_{in}(y) \equiv \phi_{in}(y)$. The change from the argument $y^\mu$ to $X^\mu(y)$ is a form of translation that can be implemented using the (non-unitary) exponentiated momentum operator as we did in eq. 5.70:

$$\phi_a(y) \equiv \Phi_a(y) = W_a^{-1}(y)\phi_a(X(y))W_a(y)$$

for a = "in" or "out." *The benefit from this approach is a clean simple definition of asymptotic particle states of definite momentum (and spin etc.).* We will follow the same strategy in two-tier QED.

*Fermion In-states and Out-States*

In this section we will develop properties of fermion in-fields and out-fields. The LSZ procedure can be schematized as:

$$\psi_{in}(y) \Rightarrow \psi_{in}(X(y)) \Rightarrow \psi(X(y)) \Rightarrow \psi_{out}(X(y)) \Rightarrow \psi_{out}(y) \qquad (7.52)$$

In-states are constructed using $\psi_{in}(y)$ which is then transformed into $\psi_{in}(X(y))$ in order to make contact with our Lagrangian formalism. The interacting field $\psi(X(y))$ is related to $\psi_{in}(X(y))$ using the standard LSZ limiting ($y^0 \to -\infty$) process. Similarly out-states are constructed using $\psi_{out}(y)$ which is transformed into $\psi_{out}(X(y))$. Again the interacting field $\psi(X(y))$ is related to $\psi_{out}(X(y))$ as part of the familiar LSZ limiting ($y^0 \to +\infty$) process.

Since much of the development differs only trivially from the standard treatment in textbooks we will simply list relevant equations and let the reader pursue them further in quantum field theory introductory textbooks.

*$\psi$ In-Field*

In order to define a perturbation theory for particle scattering we will use a free Dirac field $\psi_{in}(y)$ that satisfies

$$(i\nabla\!\!\!\!/_y - m)\psi_{in}(y) = 0 \qquad (7.53)$$

where

$$\not{\nabla}_y = \gamma^\nu \partial/\partial y^\nu$$

Defining the fourier expansion for the "bare" and "cloaked" fermion fields:

$$\psi_{in}(y) = \sum_{\pm s} \int d^3p\, N^d{}_m(p)\, [b_{in}(p,s)u(p,s)\, e^{-ip\cdot y} + d_{in}{}^\dagger(p,s)v(p,s)\, e^{ip\cdot y}] \quad (7.54a)$$

$$\psi_{in}(X(y)) = \sum_{\pm s} \int d^3p\, N^d{}_m(p)\, [b_{in}(p,s)u(p,s){:}e^{-ip\cdot X(y)}{:} + d_{in}{}^\dagger(p,s)v(p,s){:}e^{ip\cdot X(y)}{:}] \quad (7.54b)$$

we can define $\psi_{in}$ in-field states with expressions like

$$|(p_n s_n), \ldots, (p_1 s_1); (\bar{p}_m \bar{s}_m), \ldots (\bar{p}_1 \bar{s}_1)\ in> = b_{in}{}^\dagger(p_n s_n) \ldots b_{in}{}^\dagger(p_1 s_1)$$
$$d_{in}{}^\dagger(\bar{p}_m \bar{s}_m) \ldots d_{in}{}^\dagger(\bar{p}_1 \bar{s}_1)|0> \quad (7.55)$$

for n electrons and m positrons. The development parallels the conventional development of fermion in-states.

*$\psi$ Out-Field*

Similarly we can define fermion out-states for the free field $\psi_{out}(y)$. Again we will use a free Dirac field $\psi_{out}(y)$ that satisfies

$$(i\not{\nabla}_y - m)\psi_{out}(y) = 0 \quad (7.56)$$

Defining the fourier expansions for the "bare" and "cloaked" fermion fields:

$$\psi_{out}(y) = \sum_{\pm s} \int d^3p\, N^d{}_m(p)\, [b_{out}(p,s)u(p,s)\, e^{-ip\cdot y} + d_{out}{}^\dagger(p,s)v(p,s)\, e^{ip\cdot y}] \quad (7.57a)$$

$$\psi_{out}(X(y)) = \sum_{\pm s} \int d^3p\, N^d{}_m(p)\, [b_{out}(p,s)u(p,s){:}e^{-ip\cdot X(y)}{:} + d_{out}{}^\dagger(p,s)v(p,s){:}e^{ip\cdot X(y)}{:}]$$
$$(7.57b)$$

we can define bare $\psi_{out}$ out-field states with expressions like

$$|(p_n s_n), \ldots, (p_1 s_1); (\bar{p}_m \bar{s}_m), \ldots (\bar{p}_1 \bar{s}_1)\ out> = b_{out}{}^\dagger(p_n s_n) \ldots b_{out}{}^\dagger(p_1 s_1)\,\cdot$$
$$\cdot d_{out}{}^\dagger(\bar{p}_m \bar{s}_m) \ldots d_{out}{}^\dagger(\bar{p}_1 \bar{s}_1)|0> \quad (7.58)$$

for n electrons and m positrons. The development again parallels the conventional development of fermion out-states.

*Photon In and Out States*

In this section we will develop properties of photon in-fields and out-fields. The LSZ procedure can be schematized as:

$$A_{in}^{\mu}(y) \Rightarrow A_{in}^{\mu}(X(y)) \Rightarrow A^{\mu}(X(y)) \Rightarrow A_{out}^{\mu}(X(y)) \Rightarrow A_{out}^{\mu}(y) \qquad (7.59)$$

In-states are constructed using a "bare" field $A_{in}^{\mu}(y)$ which are then transformed into $A_{in}^{\mu}(X(y))$ in order to make contact with our Lagrangian formalism. The interacting field $A^{\mu}(X(y))$ is related to $A_{in}^{\mu}(X(y))$ using the standard LSZ limiting ($y^0 \rightarrow -\infty$) process. Similarly out-states are constructed using $A_{out}^{\mu}(y)$ which are transformed into $A_{out}^{\mu}(X(y))$. Again the interacting field $A^{\mu}(X(y))$ is related to $A_{out}^{\mu}(X(y))$ as part of the familiar LSZ limiting ($y^0 \rightarrow +\infty$) process.

The fourier expansions of the free "bare" and "cloaked" photon in and out fields are

$$A_a^{i}(y) = \int d^3k\, N_0(k) \sum_{\lambda=1}^{2} \varepsilon^{i}(k, \lambda)[a_a(k,\lambda)\, e^{-ik\cdot y} + a_a^{\dagger}(k,\lambda)\, e^{ik\cdot y}] \qquad (7.60a)$$

and

$$A_a^{i}(X(y)) = \int d^3k\, N_0(k) \sum_{\lambda=1}^{2} \varepsilon^{i}(k, \lambda)[a_a(k,\lambda){:}e^{-ik\cdot X(y)}{:} + a_a^{\dagger}(k,\lambda){:}e^{ik\cdot X(y)}{:}] \qquad (7.60b)$$

where a = "in" or 'out."
We can define bare photon in-field states with expressions like

$$|(k_n\lambda_n), \ldots, (k_1\lambda_1)\ in\rangle = a_{in}^{\dagger}(k_n\lambda_n) \ldots a_{in}^{\dagger}(k_1\lambda_1)|0\rangle \qquad (7.61)$$

for n photons. The development parallels the conventional development of photon in-states as does the definition of photon out-states.

## S Matrix

The S matrix is defined in a familiar way by

$$\psi_{in}(y) = S\psi_{out}(y)S^{-1} \qquad (7.62)$$

$$A^{\mu}{}_{in}(y) = SA^{\mu}{}_{out}(y)S^{-1} \qquad (7.63)$$

and the other standard properties of the S matrix with the sole exception being the form of the unitarity relation (which was discussed in the previous chapter).

## LSZ Reduction

In this section we will determine the reduction formula for fermions and photons for the S matrix in two-tier QED.

*Dirac Fields*

Consider a charged Dirac particle such as an electron. The S matrix element corresponding to an in-state: $\beta$ plus one Dirac particle of momentum p and spin s, and an out state $\alpha$ can be represented by

$$S_{\alpha\beta_{ps}} = \langle \alpha \text{ out} | \beta \text{ (ps) in}\rangle \quad (7.64)$$

which becomes

$$S_{\alpha\beta_{ps}} = \langle \alpha - \text{(ps) out} | \beta \text{ in}\rangle + \langle \alpha \text{ out} | \int d^3y\, U_{ps}(y)[\psi_{in}^{\dagger}(y) - \psi_{out}^{\dagger}(y)] | \beta \text{ in}\rangle \quad (7.65)$$

through standard manipulations where $\langle \alpha - \text{(ps) out}|$ is an out state with a particle of momentum p and spin s removed (if present) and where

$$U_{ps}(y) = \{m/[(2\pi)^3E_p]\}^{1/2}u(p,s)e^{-ip\cdot y} \quad (7.66a)$$

and for later use

$$V_{ps}(y) = \{m/[(2\pi)^3E_p]\}^{1/2}v(p,s)e^{ip\cdot y} \quad (7.66b)$$

Eq. 7.65 can be reexpressed as

$$S_{\alpha\beta ps} = S_{\alpha-ps\beta} + \langle \alpha \text{ out} | \int d^3y U_{ps}(y) W_{QED}^{-1}[\psi_{in}^{\dagger}(X(y)) - \psi_{out}^{\dagger}(X(y))]W_{QED} | \beta \text{ in}\rangle \quad (7.67)$$

using $W_{QED}(y) = W_{QEDin}(y)$ with

$$\psi_a(y) = W_{QEDa}^{-1}(y)\psi_a(X(y))W_{QEDa}(y) \quad (7.68a)$$

$$\psi_a^{\dagger}(y) = VW_{QEDa}^{-1}(y)\psi_a^{\dagger}(X(y))W_{QEDa}(y)V \quad (7.68b)$$

where the label a = "in" or a = "out", and where

$$W_{QEDa}(y) = \exp(-\mathbf{Y}(y)\cdot\mathbf{P}_{QEDa}/M_c^2) \tag{7.69}$$

and

$$W_{QEDa}^{-1}(y) = \exp(\mathbf{Y}(y)\cdot\mathbf{P}_{QEDa}/M_c^2) \tag{7.70}$$

in the Coulomb gauge of Y with $\mathbf{P}_{QEDa}$ being the spatial momentum vector for the free fermion and photon fields defined by

$$\mathbf{P}_{QEDa} = \sum_{\pm s} \int d^3p\, \mathbf{p}\, [b_a^\dagger(p,s)b_a(p,s) + d_a^\dagger(p,s)d_a(p,s)] + \sum_{\lambda=1}^{2} \int d^3k\, \mathbf{k}\, a_a^\dagger(k,\lambda)a_a(k,\lambda) \tag{7.71}$$

using the free fermion and photon creation and annihilation operators.

We note that $W_{QEDa}(y)$ is not a unitary operator – a similar situation to that of the scalar particle quantum field theory – but is pseudo-unitary:

$$W_{QEDa}^{-1}(y) = V\, W_{QEDa}^\dagger(y)\, V^{-1} \tag{7.72}$$

where (letting $a_Y^\dagger(k, \lambda)$ and $a_Y(k, \lambda)$ represent the creation and annihilation operators of the Y field) V is given by

$$V = \exp(-i\pi \sum_{\lambda=1}^{2} \int d^3k\, a_Y^\dagger(k, \lambda)a_Y(k, \lambda)) \tag{7.73}$$

V is a unitary operator with the property

$$V\, Y^j(y)\, V^{-1} = -Y^j(y) \tag{7.74}$$

for j = 1,2,3. We note (as in the scalar field discussion)

$$V^\dagger = V^{-1} = V \tag{7.75}$$

and thus

$$V^2 = I \tag{7.76}$$

V is a metric operator in the sense of Dirac as discussed earlier. We also note:

$$X^\mu(y) = V[X^\mu(y)]^\dagger V^{-1} \tag{7.77}$$

$$\psi_a^\dagger(X(y)) = V[\psi_a(X(y))]^\dagger V^{-1} \tag{7.78}$$

$$A^\mu_a(X(y)) = V[A^\mu_a(X(y))]^\dagger V^{-1} \tag{7.79}$$

for a = "in" or "out." These properties are required for eqs. 7.68a and 7.68b to hold.

The interacting $\psi(X(y))$ field approaches the in and out fields $\psi_{in}(X(y))$ and $\psi_{out}(X(y))$ in the limit that $y^0 \to -\infty$ and $y^0 \to +\infty$ respectively in the sense of Lehmann, Symanzik and Zimmermann which we *symbolize* as:

$$\psi(X(y)) \to \sqrt{Z_2}\, \psi_{in}(X(y)) \quad \text{as} \quad y^0 \to -\infty \tag{7.72}$$

$$\psi(X(y)) \to \sqrt{Z_2}\, \psi_{out}(X(y)) \quad \text{as} \quad y^0 \to +\infty \tag{7.73}$$

with $\sqrt{Z_2}$ the wave function renormalization constant. Using eqs. 7.72 and 7.73 and following the standard LSZ reduction procedure leads to:

$$S_{\alpha\beta ps} = S_{\alpha-ps\beta} - iZ_2^{-1/2}\int d^4y \langle\alpha\ \text{out}|W_{QED}^{-1}\bar{\psi}(X(y))W_{QED}|\beta\ \text{in}\rangle(-i\overleftarrow{\not\nabla}_y - m)U_{ps}(y) \tag{7.74}$$

Eq. 7.74 is similar to the usual LSZ reduction formula for a fermion extracted from an in-state except for the appearance of the W(y) operator and its inverse. We note that $W(y) = W_{in}(y)$ still because $\mathbf{P}_{QEDin}$ is independent of $y^0$.

The expressions for the other possible reductions of a fermion and its anti-particle are:

1. Reduction of an anti-particle from an in-state

$$iZ_2^{-1/2}\int d^4y \bar{V}_{\bar{p}\bar{s}}(y)(i\not\nabla_y - m)\langle\alpha\ \text{out}|W_{QED}^{-1}\psi(X(y))W_{QED}|\beta\ \text{in}\rangle \tag{7.75}$$

2. Reduction of a particle from an out-state

$$-iZ_2^{-1/2}\int d^4y \bar{U}_{p's'}(y)(i\not\nabla_y - m)\langle\alpha\ \text{out}|W_{QED}^{-1}\psi(X(y))W_{QED}|\beta\ \text{in}\rangle \tag{7.76}$$

3. Reduction of an anti-particle from an out-state

$$iZ_2^{-1/2}\int d^4y <\alpha \text{ out}|W_{QED}^{-1}\bar{\psi}(X(y))W_{QED}|\beta \text{ in}>(-i\overleftarrow{\nabla}_y - m)V_{\bar{p}'\bar{s}'}(y) \qquad (7.77)$$

where

$$V_{ps}(y) = \{m/[(2\pi)^3E_p]\}^{1/2}v(p,s)e^{ip\cdot y} \qquad (7.78)$$

*Electromagnetic Field*

The LSZ reduction of a photon from an S matrix element:

$$S_{\alpha\beta k\lambda} = <\alpha \text{ out}|\beta\ \gamma(k\lambda) \text{ in}> \qquad (7.79)$$

begins with

$$S_{\alpha\beta k\lambda} = <\alpha - \gamma(k\lambda) \text{ out}|\beta \text{ in}>$$
$$- iZ_3^{-1/2}\int d^3y\ A_{k\lambda}^{\mu *}(y)<\alpha \text{ out}|[A_{in}^{\mu}(y) - A_{out}^{\mu}(y)]|\beta \text{ in}> \qquad (7.80)$$

where $Z_3$ is a normalization constant and

$$A_{k\lambda}^{\mu}(y) = [(2\pi)^3\omega_k]^{-1/2}e^{-ik\cdot y}\ \varepsilon^{\mu}(k, \lambda) \qquad (7.81)$$

Using the LSZ symbolic notation we see

$$A^{\mu}(X(y)) \rightarrow \surd Z_3\ A_{in}^{\mu}(X(y)) \quad \text{as} \quad y^0 \rightarrow -\infty \qquad (7.82)$$

$$A^{\mu}(X(y)) \rightarrow \surd Z_3\ A_{out}^{\mu}(X(y)) \quad \text{as} \quad y^0 \rightarrow \infty \qquad (7.83)$$

and

$$A_a^{\mu}(y) = W_{QEDa}^{-1}(y)A_a^{\mu}(X(y))W_{QEDa}(y) \qquad (7.84)$$

where a = "in" or "out". Next we arrive at the reduction expression following steps that parallel the scalar field reduction:

$$S_{\alpha\beta k\lambda} = <\alpha - \gamma(k\lambda) \text{ out}|\beta \text{ in}>$$
$$- iZ_3^{-1/2}\int d^4y\ A_{k\lambda}^{\mu *}(y)\Box_y<\alpha \text{ out}|W_{QED}^{-1}A^{\mu}(X(y))W_{QED}|\beta \text{ in}> \qquad (7.85)$$

## Time Ordered Products and Perturbation Theory

By repeated application of the LSZ procedure outlined above, an S matrix element is reduced to the vacuum expectation value of time ordered products of fermion and photon fields.

$$<\alpha \text{ out}|\beta \text{ in}> = \ldots <0|T(\ldots W^{-1}(y_m)U^{-1}(y_m{}^0)\psi_{in}(X(y_m))U(y_m{}^0)W(y_m) \ldots$$
$$W^{-1}(y_n)U^{-1}(y_n{}^0)\bar{\psi}_{in}(X(y_n))U(y_n{}^0)W(y_n) \ldots$$
$$W^{-1}(y_p)U^{-1}(y_p{}^0)A_{in}{}^{\mu}(X(y_p))U(y_p{}^0)W(y_p) \ldots)|0> \ldots \tag{7.86}$$

Following the same development of the U matrix as described in chapter 6 with minor changes in details leads to the time ordered product for S matrix elements:

$$\tau(y_1,\ldots, y_n) = \frac{<0|T(\ldots\psi_{in}(X(y_m))\ldots\bar{\psi}_{in}(X(y_n))\ldots A_{in}{}^{\mu}(X(y_p))\ldots\exp[-i\int d^4y' \mathscr{H}_{FintQED}])|0>}{<0|T(\exp[-i\int d^4y' \mathscr{H}_{FintQED}])|0>} \tag{7.87}$$

where the QED interaction hamiltonian is

$$\mathscr{H}_{FintQED} = e_0:\bar{\psi}_{in}(X(y))\not{A}_{in}(X(y))\psi_{in}(X(y)): \tag{7.88}$$

plus mass counter terms.

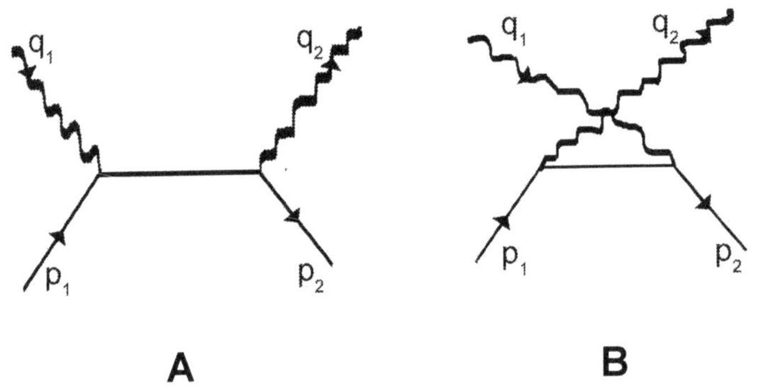


Figure 7.1. Lowest order elastic photon-electron scattering diagrams.

*Example - Photon-Electron Elastic Scattering*

In order to illustrate perturbative calculations in two-tier QED we will calculate the lowest order photon-electron elastic scattering S matrix element (Fig. 7.1).

We will see that at large distances (where the momenta are $\ll M_c$) the result is the same as the conventional QED calculation. At short distances (where the momenta are $\gg M_c$) the result differs markedly due to the effects of the Y field. The S matrix element containing the contribution of these diagrams is

$$S_{\gamma e} = (Z_2 Z_3)^{-1} e_0^{\,2} \prod_{j=1}^{6} \int d^4 y_j \, \bar{U}_{p_2 s_2}(y_4) A_{k_1\lambda_1}{}^{\mu_1}(y_2) A_{k_2\lambda_2}{}^{\mu_2 *}(y_3) \Box_{y_2} \Box_{y_3} \cdot$$

$$\cdot (i\overrightarrow{\nabla\!\!\!/}_{y_4} - m) <0| T(\bar{\psi}_{in}(X(y_1)) \psi_{in}(X(y_4)) A_{in}{}^{\mu_1}(X(y_2)) A_{in}{}^{\mu_2}(X(y_3)) \cdot$$

$$\cdot :\bar{\psi}_{in}(X(y_5)) A\!\!\!/_{in}(X(y_5)) \psi_{in}(X(y_5)) :: \bar{\psi}_{in}(X(y_6)) A\!\!\!/_{in}(X(y_6)) \psi_{in}(X(y_6)) :$$

$$) |0> (\overleftarrow{-i\nabla\!\!\!/_{y_1}} - m) U_{p_1 s_1}(y_1) \tag{7.89}$$

Diagram A in Fig. 7.1 corresponds to the Wick expansion:

$$S_{\gamma eA} = (Z_2 Z_3)^{-1} e_0^{\,2} \prod_{j=1}^{6} \int d^4 y_j \, A_{k_1\lambda_1}{}^{\mu_1}(y_2) A_{k_2\lambda_2}{}^{\mu_2 *}(y_3) \bar{U}_{p_2 s_2}(y_4) \{$$

$$(i\overrightarrow{\nabla\!\!\!/}_{y_4} - m) iS_F{}^{TT}(y_4 - y_6) \gamma^{\nu} iS_F{}^{TT}(y_6 - y_5) \gamma^{\mu} iS_F{}^{TT}(y_5 - y_1) (\overleftarrow{-i\nabla\!\!\!/_{y_1}} - m) \}$$

$$U_{p_1 s_1}(y_1) \Box_{y_2} \Box_{y_3} \, iD_F{}^{trTT}(y_5 - y_2)_{\mu_1 \mu} iD_F{}^{trTT}(y_3 - y_6)_{\mu_2 \nu} \tag{7.90}$$

After some manipulation eq. 7.90 can be placed in the form:

$$S_{\gamma eA} = (Z_2 Z_3)^{-1} e_0^{\,2} (2\pi)^4 \delta^4(p_2 + k_2 - p_1 - k_1) \bar{u}(p_2, s_2) \mathscr{S}_L{}^{TT}(p_2) \, \epsilon\!\!/(k_2, \lambda_2) \cdot$$

$$\cdot \mathscr{S}^{TT}(p_1 + k_1) \, \epsilon\!\!/(k_1, \lambda_1) \mathscr{S}_R{}^{TT}(p_1) u(p_1, s_1) \mathscr{D}^{TT}(k_1) \mathscr{D}^{TT}(k_2) \tag{7.91}$$

where

$$\mathscr{S}_L^{\,TT}(p) = iN_{fp} \int d^4y\, e^{ip\cdot y}\, (i\nabla\!\!\!/_y - m)\, S_F^{\,TT}(y) \tag{7.92}$$

$$\mathscr{S}^{\,TT}(p) = i \int d^4y\, e^{ip\cdot y}\, S_F^{\,TT}(y) \tag{7.93}$$

$$\mathscr{S}_R^{\,TT}(p) = iN_{fp} \int d^4y\, e^{ip\cdot y}\, S_F^{\,TT}(y)\, \overleftarrow{(i\nabla\!\!\!/_y - m)} \tag{7.94}$$

$$\mathscr{D}^{\,TT}(k) = iN_{\gamma k} \int d^4y\, e^{ik\cdot y}\, \Box_y\, D_F^{\,trTT}(y)_{00} \tag{7.95}$$

with

$$N_{fp} = \{m/[(2\pi)^3 E_p]\}^{1/2} \tag{7.96}$$

and

$$N_{\gamma k} = [(2\pi)^3 \omega_k]^{-1/2} \tag{7.97}$$

The factors $\mathscr{S}_L^{\,TT}(p_2)$, $\mathscr{S}_R^{\,TT}(p_1)$, and $\mathscr{D}^{\,TT}(k_1)$ and $\mathscr{D}^{\,TT}(k_2)$ serve to normalize the in and out particle legs. They are a consequence of the dressing of the legs by Y particle "clouds." At long distance (low momentum) they approach the corresponding values of conventional QED:

$$\mathscr{S}_L^{\,TT}(p) \to iN_{fp} \tag{7.98}$$

$$\mathscr{S}_R^{\,TT}(p) \to iN_{fp} \tag{7.99}$$

$$\mathscr{D}^{\,TT}(k) \to iN_{\gamma k} \tag{7.100}$$

if $p, k \ll M_c$.

*Here again we must "lop off" external legs as in eqs. 6.74-6.75 to achieve a theory satisfying unitarity. We leave this as an exercise for the reader.*

*Deep e-p Inelastic Scattering Partons?*

The form of eq. 7.93 shows the fermion propagator factor $\mathscr{S}^{\,TT}(p)$ is not the simple form of a free fermion propagator. Rather it consists of a fermion traveling within a "stream" of free Y quanta. This picture is reminiscent of Feynman's picture of deep inelastic e-p scattering in which the proton is viewed as a stream of partons. This question will be addressed in a future publication.

*External Leg "Normalizations"*

The external leg factors $\mathscr{S}_L^{TT}(p)$, $\mathscr{S}_R^{TT}(p)$ and $\mathscr{D}^{TT}(k)$ change the normalization of the external legs due to the Y particle "cloud" surrounding each particle. W*e must "lop off" external legs as in eqs. 6.74-6.75 to achieve a theory satisfying unitarity. We leave this as an exercise for the reader. In this section we examine external leg factors to see the effect of the Y quanta cloud around particles.*

In order find their form for large momenta we will first evaluate the large momentum limit of the fermion propagator as $z^2 \to 0$ (the light cone). Starting from eq. 7.45 one can show that (space-like limit)

$$iS_F^{TT}(z) \to \gamma^0 \epsilon(z^0) M_c^3 [-M_c^2 \pi z^2/2]^{3/2} + \mathscr{O}(z^5) \tag{7.101}$$

where $z^2 = z_0^2 - \mathbf{z}^2$. Therefore on dimensional grounds we see that

$$\mathscr{S}^{TT}(p) \backsim \gamma^0 M_c^6 p^{-7} + \mathscr{O}(p^{-9}) \tag{7.102}$$

as p gets very large ($p \gg M_c$). At low momenta ($p \ll M_c$) the standard momentum space form of the fermion propagator is found:

$$\mathscr{S}^{TT}(p) \backsim (\not{p} - m + i\epsilon)^{-1} \tag{7.103}$$

Similarly we find

$$\mathscr{S}_L^{TT}(p) \backsim M_c^6 p^{-6} + \mathscr{O}(p^{-8}) \tag{7.104}$$

$$\mathscr{S}_R^{TT}(p) \backsim M_c^6 p^{-6} + \mathscr{O}(p^{-8}) \tag{7.105}$$

as p gets very large ($p \gg M_c$), and

$$\mathscr{D}^{TT}(k) \backsim M_c^4 k^{-4} \tag{7.106}$$

from eqs. 7.47 and 4.24 as k gets very large ($k \gg M_c$).

For small momenta compared to $M_c$ we find the usual normalization:

$$\mathscr{S}_L^{TT}(p) = \mathscr{S}_R^{TT}(p) = iN_{fp} \tag{7.107}$$

$$\mathscr{D}^{TT}(k) = iN_{\gamma k} \qquad (7.108)$$

Thus for low momenta (p, k $\ll M_c$) $S_{\gamma eA}$ yields the standard result of QED while at large momenta (p, k $\gg M_c$) we see a high power of inverse momentum showing the well behaved nature of the theory at short distances.

## Renormalization of Two-Tier QED

Two-tier QED is a finite quantum field theory satisfying the unitarity condition. The degree of divergence of a Feynman diagram term in *conventional QED* is

$$D = 4k - 2b - f \qquad (7.109)$$

where

k = the number of internal momentum integrations
b = the number of internal photon lines
f = the number of internal electron lines

Many diagrams are thus divergent in conventional QED and a renormalization program must be followed to achieve a theory with all divergences formally absorbed into renormalizations of the fundamental parameters of the theory. Despite the success of this approach and the excellent agreement of QED with experiment the presence of divergences in QED is logically unsatisfactory and suggests QED, and its successors Electroweak Theory and the Standard Model, are at best interim theories. Numerous attempts have been made to modify QED in order to eliminate its divergences. Some noteworthy attempts include the Lee-Wick formulation of QED and the Johnson-Baker-Willey model of QED. None of these attempts have succeeded for one reason or another.[35]

The degree of divergence formula is different in two-tier QED. It demonstrates there are no divergences in two-tier QED. Thus it would be more aptly named the "degree of convergence." The formula is:

$$D^{TT} = 4k - 6b - 7f \qquad (7.110)$$

with k, b and f as above. The coefficient of b is 6 in two-tier QED because the two-tier photon propagator behaves as $k^{-6}$ at high momentum (See eq. 7.47 for the photon

---

[35] Lee, T. D. and Wick, G. C., Phys. Rev. **D2**, 1033 (1970); T. D. Lee and G. C. Wick, Nucl. Phys. **B9**, 209 (1969) and references therein; M. Baker and K. Johnson, Phys. Rev. **D3**, 2516 (1971); M. Baker and K. Johnson, Phys. Rev. **D8**, 1110 (1973); K. Johnson, M. Baker, and R. Willey, Phys. Rev. **136**, B1111 (1964); K. Johnson, R. Willey, and M. Baker, Phys. Rev. **163**, 1699 (1967); S. Blaha,, Phys. Rev. **D9**, 2246 (1974) and references therein; S. Adler, Phys. Rev. **D5**, 3021 (1972).

propagator. Eq. 4.25 shows its high momentum behavior). The coefficient of f is 7 in two-tier QED because the two-tier fermion propagator behaves as $k^{-7}$ at high momentum (see eq. 7.102).

*We note the degree of Divergence in Two-Tier QED is always negative – indicating a finite theory!*

$$D^{TT} < 0 \tag{7.111}$$

For example the degree of divergence in two-tier QED of the lowest order in $\alpha$ (see Fig. 7.2): i) vacuum polarization diagram is $D^{TT} = -10$, ii) fermion self-energy is $D^{TT} = -9$ and iii) electromagnetic vertex correction is $D^{TT} = -16$.

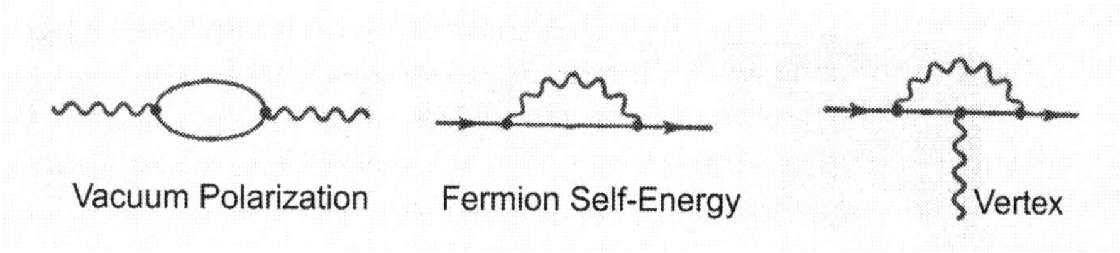


Figure 7.2 Some low order (normally divergent) diagrams for the vacuum polarization, fermion self-energy and electromagnetic vertex correction.

It is easy to see that all two-tier QED Feynman diagrams are ultra-violet finite.

## Unitarity of Two-Tier QED

The remaining major issue is unitarity. Two-tier QED satisfies the unitarity condition between physical states. The argument demonstrating unitarity parallels the discussion of unitarity for scalar $\phi^4$ quantum field theory in chapter 6.

Two-tier QED *superficially* appears to have a unitarity problem due to the non-hermitean nature of its hamiltonian. The lack of hermiticity is entirely due to the appearance of $iY^\mu$ in the $X^\mu$ field coordinates. Thus the interaction hamiltonian in eq. 7.88 is not hermitean:

$$H_{FintQED} = \int d^3y' \mathscr{H}_{FintQED}(\bar{\psi}_{in}(y' + iY(y')/M_c^2), A_{in}^{\ \mu}(y' + iY(y')/M_c^2), \psi_{in}(y' + iY(y')/M_c^2)) \tag{7.112}$$

and

$$H_{FintQED} \neq H_{FintQED}^{\ \dagger} = \int d^3y' \mathscr{H}_{FintQED}(\bar{\psi}_{in}(y' - iY(y')/M_c^2), A_{in}^{\ \mu}(y' - iY(y')/M_c^2), \psi_{in}(y' - iY(y')/M_c^2)) \tag{7.113}$$

The metric operator (eq. 7.73) establishes the relation between $H_{FintQED}$ and its hermitean conjugate is

$$H_{FintQED} = V\, H_{FintQED}^{\dagger}\, V \qquad (7.114)$$

Thus the Two-Tier QED S matrix is not unitary – S is pseudo-unitary in general:

$$S^{-1} = V\, S^{\dagger}\, V \qquad (7.115)$$

which implies

$$VS^{\dagger}VS = I \qquad (7.116)$$

*Two-Tier QED S matrix satisfies unitarity between physical asymptotic states* which in two-tier QED are states consisting of charged fermions, and photons. The proof is identical to eqs. 6.46 – 6.48.

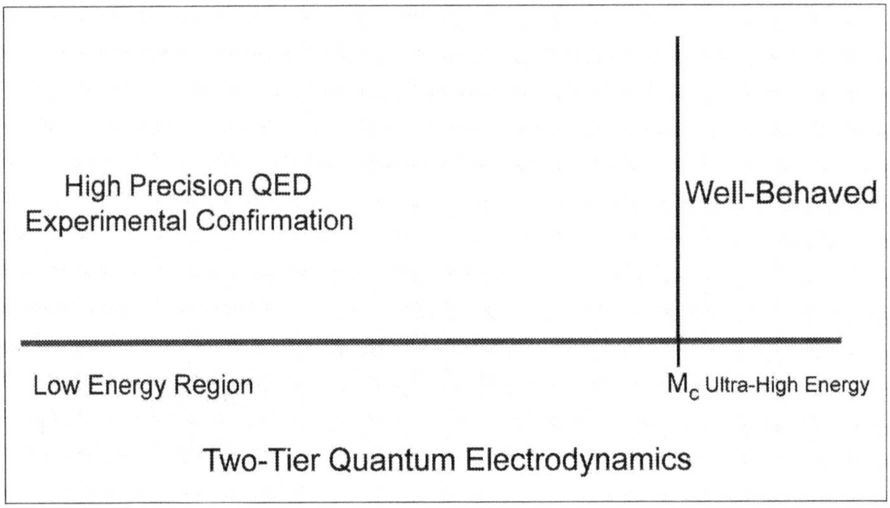


Figure 7.3. A Schematic View of Two-Tier QED.

## Correspondence Principle for Two-Tier QED

The two-tier QED theory that we have developed has the behavior of conventional QED at "low energies." Since $M_c$ sets the scale at which deviations from the normal QED results become significant we can set $M_c$ to an extremely high value such as the Planck mass and obtain as close an agreement with conventional QED as we wish. Thus two-tier QED implements a type of Correspondence Principle with conventional QED as its "low energy" limit.

## Consistency with Precision QED Experiments

Since we can obtain results as close to conventional QED with a sufficiently high choice of $M_c$, two-tier QED has the same excellent agreement[36] with experiment as conventional QED.

## Calculational Rules for Feynman Diagrams

This section describes the procedure for the calculation of S matrix terms corresponding Feynman diagrams in Two-Tier QED and other quantum field theories. By design the approach parallels the perturbative approaches of conventional QED and other quantum field theories. The examples considered earlier in this chapter, and the preceding chapter, illustrate two-tier perturbation theory.

Procedure: Follow the conventional procedure to form the expression for each Feynman diagram in the perturbative calculation. Then replace each propagator in each expression with the corresponding two-tier propagator as specified below. Then evaluate the expression.

*Feynman Propagators and their Two-Tier Propagator Equivalents*

Two-tier propagators are denoted with the superscript "TT".

Spin 0 Propagator Case

$$\Delta_F(p) = (p^2 - m^2 + i\varepsilon)^{-1} \quad \rightarrow \quad \int d^4z\, e^{+ip\cdot z}\, \Delta_F^{TT}(z) \qquad (7.120)$$

where $\Delta_F^{TT}(z)$ is given by eq. 4.9.

Spin 1 Photon Propagator Case

$$D_F(p)_{\mu\nu} = -g_{\mu\nu}(p^2 + i\varepsilon)^{-1} \quad \rightarrow \quad \int d^4z\, e^{+ip\cdot z}\, D_F^{TT}(z)_{\mu\nu} \qquad (7.121)$$

where $D_F^{TT}(z)_{\mu\nu}$ is given by eq. 7.47.

Spin ½ Fermion Propagator Case

$$S_F(p) = (\not{p} - m + i\varepsilon)^{-1} \quad \rightarrow \quad \int d^4z\, e^{+ip\cdot z}\, S_F^{TT}(z) \qquad (7.122)$$

---

[36] See, for example, T. Kinoshita, "The Fine Structure Constant", Cornell University preprint CLNS 96/1406 (1996); V. W. Hughes and T. Kinoshita, Rev. Mod. Phys. **71**, S133 (1999) and references therein.

where $S_F^{TT}(z)$ is given by eq. 6.88.

Spin 2 Massless Boson (graviton) Propagator Case

$$\Delta_{F2}(p)_{\mu\nu\rho\sigma} = b_{\mu\nu\rho\sigma}(p^2 - m^2 + i\varepsilon)^{-1} \quad \rightarrow \quad \int d^4z\, e^{+ip\cdot z}\, \Delta_{F2}^{TT}(z)_{\mu\nu\rho\sigma} \qquad (7.123)$$

where $\Delta_{F2}^{TT}(z)_{\mu\nu\rho\sigma}$ the graviton propagator.

## Two-Tier Coulomb Potential vs. Conventional Coulomb Potential

The familiar Coulomb potential is (for two particles of opposite unit electric charge):

$$V_{\text{Coulomb}} = -\alpha/|\mathbf{r}| \qquad (7.124)$$

The Two-Tier QED Coulomb potential (eq. 7.49) is:

$$V_{\text{Two-TierCoul}} = -\alpha\Phi(M_c^2\pi|\mathbf{r}|^2)/|\mathbf{r}| \qquad (7.125)$$

where $\Phi(x)$ is the error function.[37] At small distances ($\pi r^2 \ll M_c^{-2}$)

$$V_{\text{Two-TierCoul}} \rightarrow -2\alpha\sqrt{\pi}\, M_c^2|\mathbf{r}| \qquad (7.126)$$

a linear potential, and at large distances ($\pi r^2 \gg M_c^{-2}$)

$$V_{\text{Two-TierCoul}} \rightarrow V_{\text{Coulomb}} \qquad (7.127)$$

The Two-Tier Coulomb potential has a minimum at

$$M_c^2\pi|\mathbf{r}|^2 = 1 \qquad (7.128)$$

37 W. Magnus and F. Oberhettinger, *Formulas and Theorems for the Special Functions of Mathematical Physics* (Chelsea Publishing Co., New York, 1949) page 96.

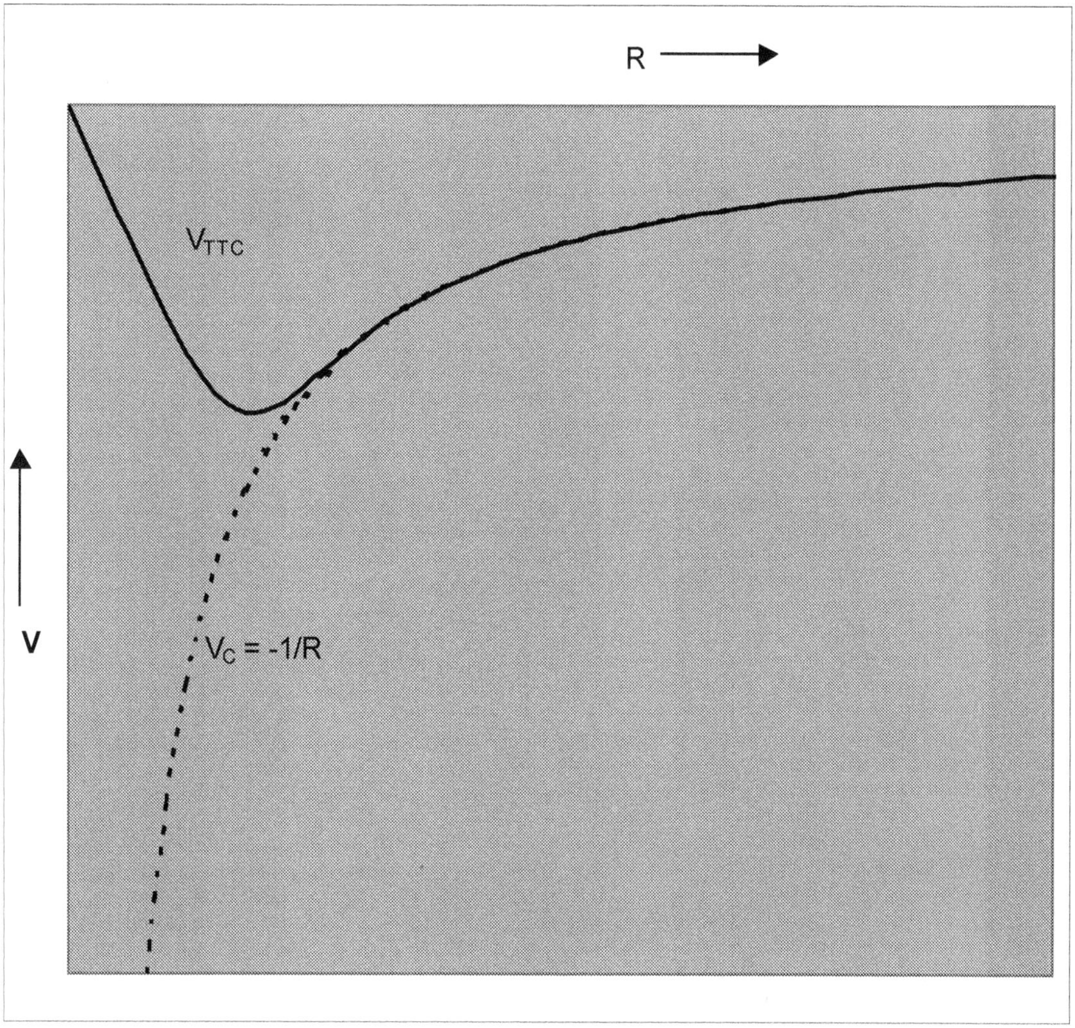


Figure 7.4. Plot of the form of the Two-Tier Coulomb "attractive" potential between particles of opposite unit charge divided by $\alpha M_c$: $V_{TTC} = V_{Two\text{-}TierCoul}/(\alpha M_c)$. $V_{TTC}$ is dimensionless. The dotted line is the conventional Coulomb attractive potential divided by $\alpha M_c$: $V_c = V_{Coulomb}/(\alpha M_c) = 1/R$. Note the Two-Tier Coulomb force between particles of opposite charge is repulsive at short distance.

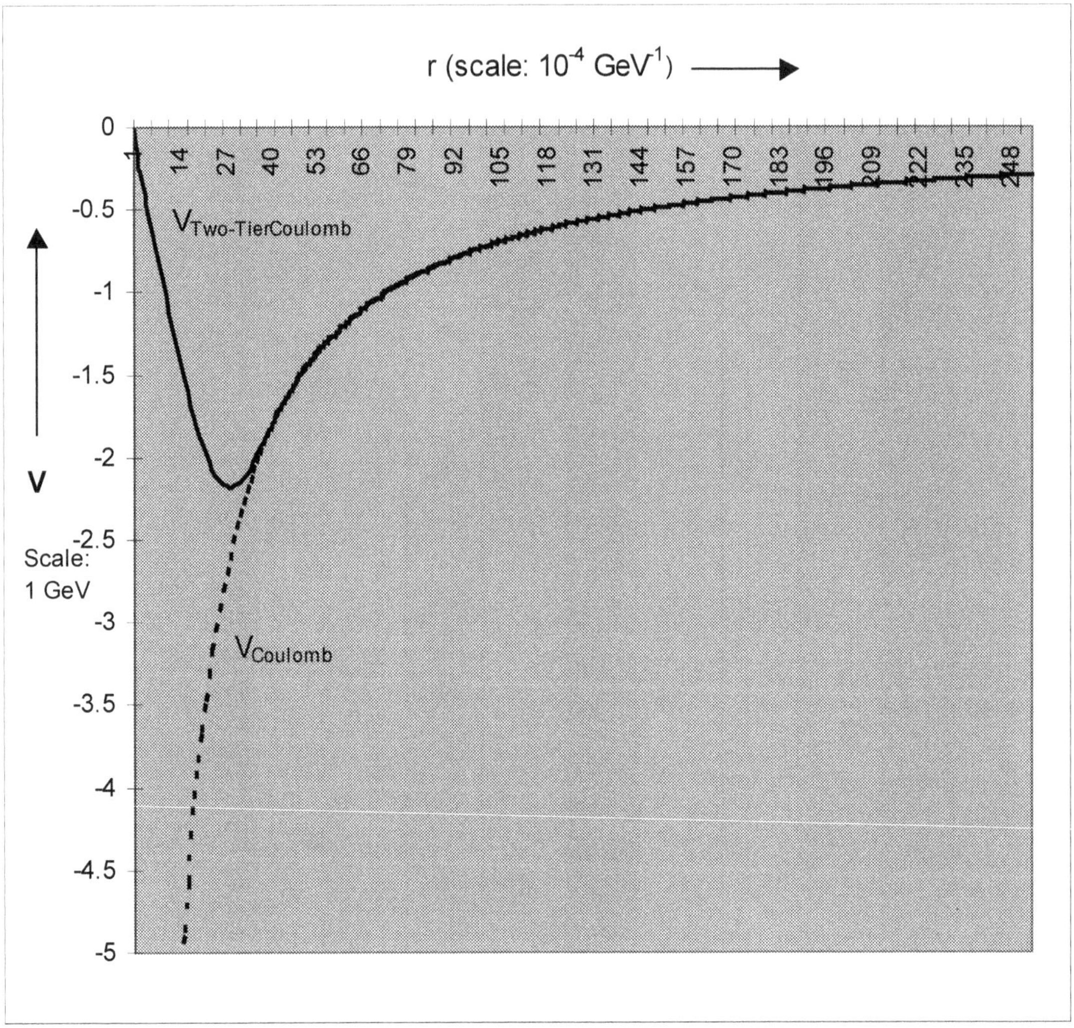


Figure 7.5. Two-Tier Coulomb Potential compared to conventional Coulomb potential for $M_c$ = 200 GeV/$c^2$. Radial distance is measured in units of $10^{-4}$ $GeV^{-1}$. The potential energy for two opposite unit charges is measured in GeV units.

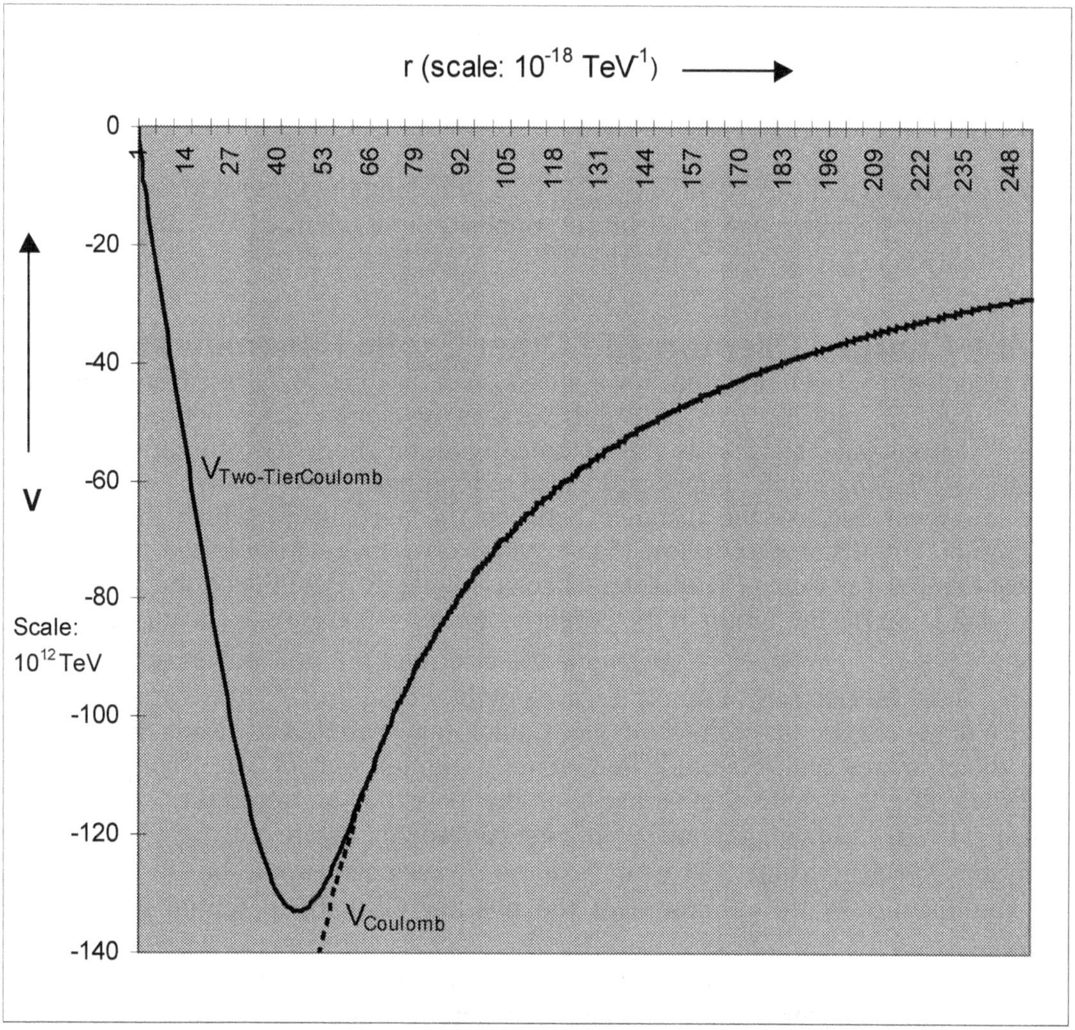


Figure 7.6. Two-Tier Coulomb Potential compared to conventional Coulomb potential for $M_c = M_{planck} = 1.22 \times 10^{19}$ GeV/c$^2$. Radial distance is in units of $10^{-18}$ TeV$^{-1}$. The potential energy for two opposite unit charges is measured in units of $10^{12}$ TeV.

At the minimum $V_{Two\text{-}TierCoul}$ has the value:

$$V_{Two\text{-}TierCoulMIN} = -\ .8427\alpha\surd\pi\ M_c \tag{7.129}$$

If we define distance in terms of the mass scale

$$|\mathbf{r}| = R/M_c \tag{7.130}$$

then

$$V_{Two\text{-}TierCoul} = -\ \alpha M_c \Phi(\pi R^2)/R \tag{7.131}$$

Fig. 7.4 displays a plot of $V_{TTC} = V_{Two\text{-}TierCoul}/(\alpha M_c)$ showing the general form of the Two-Tier Coulomb potential for particles of opposite unit charge. We also plot $V_C = V_{Coulomb}/(\alpha M_c) = 1/R$.

## Doubly-Charged Dilepton and Other Exotic Resonances

Doubly charged dilepton resonances such as $e^-e^-$ or $\mu^-\mu^-$ resonances are possible since it appears that the repulsive electromagnetic force between similarly charged leptons goes to zero as the distance between them goes to zero creating a potential barrier that might, in principle, lead to binding.

However because the distance between the particles at which the repulsive Coulomb potential starts to decline (Fig. 7.5) is very small, a very broad resonance is the best that one can expect. Tunneling will cause rapid decay of the bound state.

Note that the minimum of the potential, for $M_c = 200$ GeV/c$^2$, is at $r = 2.8 \times 10^{-5}$ GeV$^{-1}$ $\equiv 5.5 \times 10^{-19}$ cm (by eq. 7.128) while the electron Compton wavelength is $3.861 \times 10^{-13}$ m. Thus the electron's Compton wavelength is roughly $10^6$ times larger than the distance of the maximum of the Two-Tier Coulomb potential. As a result the electrons in a dilepton resonance would immediately tunnel out of the binding region (*zitterbewegung*!) despite the strength of the potential at minimum: 21.7 TeV/c$^2$ according to eq. 7.129. Even if one could create a resonance state of two $\tau^-$ leptons with $\tau$ Compton wavelengths of $1.135 \times 10^{-14}$ cm the $\tau$ wavelength would be $10^5$ times larger than the distance of the minimum of the two-tier Coulomb potential if $M_c = 200$ GeV/c$^2$.

If $M_c$ is much larger than 200 GeV/c$^2$ then the location of the peak of the potential is even smaller as shown in Fig. 7.6 where $M_c$ equals the Planck mass. Thus a dilepton resonance can be expected to be very broad and very short-lived if indeed it exists at all.

Other exotic resonances might be possible involving the strong force which also interchanges repulsion and attraction at very short distances.

# 8. Two-Tier Standard Model Theory

## Introduction

The Standard Model is a unified model of the electromagnetic, weak and strong interactions. It is formed from Electroweak Theory and an SU(3) color gauge theory of the strong interaction called Quantum Chromodynamics (QCD). Each part is separately renormalizable, as is the combined theory. This ad hoc theory of all known interactions except gravitation has been remarkably successful in accounting for the experimental data at energies currently accessible at accelerators. There are a number of phenomena that are not understood within the framework of the Standard Model such as dark matter and dark energy, some CP violation phenomena, and the spin dependence observed in certain experiments. There is also an appreciation of the ad hoc nature of the unification of the strong and electroweak sectors. They are "just glued together". An overall symmetry encompassing both sectors (broken as it would be) would be more reassuring and more elegant. However the theory works, and appears to work well, for most particle physics phenomena.

Superstring theory has become a seriously considered alternative to the Standard Model. It appears to offer a finite theoretical framework without divergences. It offers the hope of a rational grand unification. It is mathematically interesting and elegant. However, it has almost no experimental support. It predicts many particles that remain to be found. It requires additional dimensions that are not evidenced. It does not have a demonstrable, satisfactory "low energy limit" that resembles the Standard Model although ad hoc Superstring-like Standard Model variants have been studied. It has the problem of too many possibilities. After thirty years of effort one cannot point to a Superstring theory and say it is THE Superstring theory, and here is the method of symmetry-breaking that results in the Standard Model. Considering that far more theoretical effort has been expended on Superstring theory by more theorists then on any other theory of physics, its physical justification, its nature, and its contact with experiment is still not well understood.

In the hope of developing a sound theory that clearly makes contact with known experimental data, that has a direct connection to the Standard Model, that is finite (no divergences), and that can be directly extended to encompass gravitation, we have developed a two-tier formulation of the Standard Model and its variants. It is admittedly far more modest in its scope than Superstring theory. Yet it allows us to have a finite theory – a major goal of Superstring theory – and to address cosmological issues (Blaha(2004)) by its easy unification with two-tier quantum gravity (next chapter) – also a finite theory without divergences.

Thus we have a theory that accounts for all known interactions and yet incrementally extends known successful theories – in a significant way. With (perhaps) the exception of General Relativity, the growth of particle physics theory in the twentieth-century has been largely incremental.

The nature of the mechanism that two-tier quantum field theory uses to avoid divergences has the important feature that it does not depend on "magic cancellations", does not depend on details of the symmetries, and does not depend on details of the form of the theories. Thus all the variations and extensions of the Standard Model have an equivalent two-tier version that is finite. This feature gives the theorist the flexibility to modify and extend the two-tier Standard Model without worrying about the introduction of new divergences.

The other major advantage is that the theory is very much like the world we see at current energies – four dimensions and no large number of new particles. It leaves open the door for new interactions and new particles IF they are found at higher energies. Thus it is not a showstopper. Nor is it a prediction of a vast desert extending to the highest energies. It gives a finite, unified theory with an "Open Door" policy for the future.

## Formulation of Two-Tier Standard Models

There are many excellent texts[38] on the Standard Model. Therefore this author sees no reason to recapitulate the features of the Standard Model. We will focus on the form and features of a generic two-tier version of the Standard Model.

We will assume that the form of the Standard Model Lagrangian is:

$$\mathscr{L}_{SM} = J\mathscr{L}_F^{SM}(X^\mu) + \mathscr{L}_C(X^\mu(y), \partial X^\mu(y)/\partial y^\nu) \tag{8.1}$$

where $\mathscr{L}_F^{SM}$ is the complete "normal" quantum field theory Lagrangian for the Standard Model variant under consideration and J is a Jacobian (eq. A.21). *All particle fields in* $\mathscr{L}_F^{SM}$ *are assumed to be functions of the* $X^\mu$ *coordinate only. The dependence of the particle fields on the "underlying" coordinates* $y^\mu$ *is assumed to be solely through* $X^\mu$. The Lagrangian $\mathscr{L}_{SM}$ is a separable Lagrangian of the type of eq. A.26 embodying the composition of extrema described in Appendix A. Scalar particle examples of separable

[38] Some of the many excellent texts: P. Ramond, *Journeys Beyond the Standard Model* (Westview Press, New York, 1999); W. N. Cottingham and D. A. Greenwood, *An Introduction to the Standard Model of Particle Physics* (Cambridge University Press, New York, 1999); K. Huang, *Quarks, Leptons and Gauge Fields* (World Scientific, River Edge, NJ 1992); T. P, Cheng and L. F. Li, *Gauge Theories of the Elementary Particles* (Oxford University Press, New York, 1988); J. Donoghue, E. Golowich and B. Holstein, *Dynamics of the Standard Model* (Cambridge University Press, New York, 1994).

Lagrangians appear in eqs. A.96, 3.10 and 3.15. A scalar particle $\phi^4$ theory with a separable Lagrangian is specified by eqs. 5.20 – 5.24 and decribed in detail in chapters 5 and 6. The separable Lagrangian two-tier version of QED is specified by eqs. 7.1 – 7.5 and described in some detail in chapter 7.

In all of these cases the coordinate part of the Lagrangian $\mathscr{L}_C$ as

$$\mathscr{L}_C = -\tfrac{1}{4}\, F_Y^{\mu\nu} F_{Y\mu\nu} \tag{3.15}$$

with

$$F_{Y\mu\nu} = \partial Y_\mu/\partial y^\nu - \partial Y_\nu/\partial y^\mu \tag{3.14}$$

The field equations of the theory are found using the conventional approach. The Hamiltonian, energy-momentum tensor and conserved quantities are also found using a conventional approach as illustrated in chapters 2, 3 and 4. The canonical quantization procedure is also followed. Thus the application of these procedures to the Standard Model will yield the standard results with the sole difference being that all fields are functions of $X^\mu$ and all derivatives of fields are derivatives with respect to $X^\mu$. A perturbation theory along the lines of chapters 5 and 6 can then be directly developed.

The net result is that the free field Feynman propagators in a Feynman diagram approach to perturbative calculations are each replaced with the corresponding two-tier propagator. All coordinate space integrations are over the y coordinates.

**Rule 1:** *If the conventional coordinate space Feynman propagator for a free particle has the form:*

$$G\ldots(z) = \int d^4p\, e^{-ip\cdot z}\, G\ldots(p)/(2\pi)^4 \tag{8.2}$$

*where ... represents any space-time indices, spin indices and internal symmetry indices, then the equivalent coordinate space (y – coordinates) two-tier propagator is*

$$G\ldots^{TT}(z) = \int d^4p\, e^{-ip\cdot z}\, G\ldots(p) R(p, z)/(2\pi)^4 \tag{8.3}$$

*where R(p, z) is specified in eqs. 4.12 – 4.20.*

The vertex expressions remain identically the same as in the conventional Standard Model. All integrations are done in the y-coordinate space as shown in the perturbation discussion leading to eq. 6.37 and the perturbation theory calculations at the end of chapter 6 and in chapter 7.

*Procedure to Construct the Two-Tier Equivalent of a Conventional Feynman Diagram Expression*

After we form the expression for a Feynman diagram in a conventional Standard Model variant, we can construct the Two-Tier equivalent by taking *every* particle or ghost ghost propagator factor of the form:

$$G\ldots(p)$$

in momentum space, and replacing it with

$$G\ldots^{TT}(p) = \int d^4z\, e^{ip\cdot z}\, G\ldots^{TT}(z)$$

$$= \int d^4z\, e^{ip\cdot z} \int d^4p'\, e^{-ip'\cdot z}\, G\ldots(p')R(p', z)/(2\pi)^4$$

The resulting expression is the two-tier expression for the Feynman diagram.

## "Low Energy" Behavior of the Two-Tier Standard Model

Since $R(p, z) \cong 1$ for $p \ll M_c$ the two-tier Standard Model is virtually identical to the conventional, corresponding Standard Model variant for energies much less than $M_c$. Since $M_c$ can be very large – perhaps equal to the Planck mass or larger, the two-tier Standard Model predictions at current energies can be made arbitrarily close to the "low energy" results of the corresponding Standard Model. The Standard Model is a limiting case of the two-tier Standard Model – thus implementing a *Correspondence Principle.*

## "High Energy" Behavior of the Two-Tier Standard Model

At high energies (short distances) where p is of the order of $M_c$ or larger, R(p, q) provides a Gaussian damping factor that makes all perturbation theory calculations ultra-violet finite. This has been shown in exhaustive detail for fermions and bosons in earlier chapters.

*Massive Vector Bosons – No Divergence Problems*

The only new case is the case of massive vector bosons. The propagator has the general form:

$$D_{VM}(k)_{\mu\nu} = -(g_{\mu\nu} - k_\mu k_\nu/m^2)(k^2 - m^2 + i\varepsilon)^{-1} \qquad (8.4)$$

in momentum space in conventional quantum field theory modulo internal symmetry indices.

The two-tier quantum field theory equivalent is:

$$D_{VM}{}^{TT}(z)_{\mu\nu} = \int d^4k\, e^{-ip\cdot z}\, D_{VM}(k)_{\mu\nu} R(k, z)/(2\pi)^4 \quad (8.5)$$

in y-coordinate space. The leading momentum space behavior at high-energy of $D_{VM}{}^{TT}$ is generated by the $g_{\mu\nu}$ term (unlike the situation in conventional quantum field theory):

$$D_{VM}{}^{TT}(k)_{\mu\nu} \sim g_{\mu\nu} k^{-6} \quad (8.6)$$

as in eq. 6.101 for the massless vector boson case. Thus a Higgs mechanism is not needed to give vector bosons mass while maintaining renormalizability in the Electroweak sector.

Parenthetically we note that the two-tier version of the ghost propagators appearing in the gauge theory sectors is also handled by Rule 1. These propagators have the same high energy leading momentum behavior as massless scalar bosons, $k^{-6}$.

Thus the leading high-energy behavior of all two-tier propagators of Standard Model particles and ghosts is a high negative power of momentum. As a result all perturbative calculations are ultra-violet finite.

## Path Integral Formulation

Until now we have been viewing quantum field theory calculations in terms of conventional, Feynman diagram-based, perturbation theory. The appearance of a number of internal symmetries in the Standard Model (usually SU(2)⊗SU(1)⊗SU(3)) that are implemented with Yang-Mills gauge fields often makes it convenient to use a path integral formalism. Since there are many excellent introductions to the path integral formalism in relation to Yang-Mills fields and the Standard Model we will confine our discussion to aspects specifically related to the two-tier formalism for quantum field theory that we have been developing.

We will consider the case of a two-tier scalar particle quantum field theory initially for the sake of simplicity, and then consider two-tier Yang-Mills gauge field theories. The primary quantity in the path integral formulation is

$$Z(J) = \langle 0^+ | 0^- \rangle \quad (8.8)$$

$$= N \int D^4Y D\phi \exp\{ i\int d^4y\, [\mathscr{L} + j_\mu(y)Y^\mu(y) + J(y)\phi(X)]\} \quad (8.9)$$

$$= \langle 0|T\big(\exp\{i\int d^4y\, [\mathscr{L} + j_\mu(y)Y^\mu(y) + J(y)\phi(X)]\}\big)|0\rangle \quad (8.10)$$

with X defined by eq. 3.12. Eq. 8.9 expresses the path integral as a product of integrals over classical field points, and eq. 8.10 provides an operator formulation of the path integral. N is a normalization factor. (As earlier we use units where $\hbar = 1$.) We note that $<0^+|0^->$ is the probability amplitude that the vacuum state at $y^0 = -\infty$ transitions to the vacuum state at $y^0 = +\infty$. The external sources $j_\mu(y)$ and $J(y)$ are arbitrary c-number functions of the respective variables with the restriction: as $y^0 \equiv Y^0 \to \pm\infty$ (in the Y Coulomb gauge that we are using) both external sources approach zero:

$$\lim_{y0 \to \pm\infty} j_\mu(y) = \lim_{y0 \to \pm\infty} J(y) = 0 \tag{8.11}$$

The functional derivatives satisfy

$$\delta J(y_1)/\delta J(y_2) = \delta^4(y_1 - y_2) \tag{8.12}$$

and

$$\delta j_\mu(y_1)/\delta j^\nu(y_2) = g_{\mu\nu}\delta^4(y_1 - y_2) \tag{8.13}$$

*General Procedure*

In the general case we will assume that the Lagrangian (with external sources added) has the form:

$$\mathscr{L} = \mathcal{J}(Y, y)[\mathscr{L}_{Fint} + \sum_i \mathscr{L}_{F0i}] + \mathscr{L}_C \tag{8.14}$$

where $\mathscr{L}_{Fint}$ is the "interaction Lagrangian" that contains all non-quadratic field operator product terms, $\mathscr{L}_{F0i}$ is the "free field" Lagrangian for particle species i, $\mathscr{L}_C$ is the (free) Lagrangian for the $Y^\mu$ abelian gauge field described in detail in earlier chapters, and $\mathcal{J}(Y, y)$ is the Jacobian for the transformation of $y^\mu$ coordinates to $X^\mu$ coordinates. (We now use script $\mathcal{J}$ for the Jacobian because the sources use the symbol J.) $\mathscr{L}_{Fint}$ and $\mathscr{L}_{F0i}$ are functions of the fields corresponding to the physical particles of the theory. Each field is solely a function of $X^\mu$ and all derivatives are with respect to $X^\mu$ as developed in the two-tier models of preceding chapters:

$$X_\mu(y) = y_\mu + i\,Y_\mu(y)/M_c^2 \tag{3.12}$$

In the present discussion we will deal directly with the $Y^\mu$ field and use eq. 3.12 as a notational convenience. We begin with eqs. 8.9 and 8.14 for the case of n scalar Klein-Gordon particles, which we write as:

$$Z(J) = N \{\exp\{ i\int d^4y\, \mathcal{J}(-i\delta/\delta j^\nu(y), y)\mathcal{L}_{Fint}(-i\delta/\delta J_1(y), \ldots, -i\delta/\delta J_n(y))\}\cdot$$

$$\cdot\prod_k [ \int D\phi_k \exp\{ i\int d^4y\, [\mathcal{J}(-i\delta/\delta j^\nu(y), y)\mathcal{L}_{F0k} + J_k(y)\phi_k(X)]\} |_{j_\mu = 0}]\cdot$$

$$\cdot\int D^4Y \exp\{ i\int d^4y[\mathcal{L}_C + j_\mu(y)Y^\mu(y)]\}\} |_{j_\mu = 0} \tag{8.15}$$

using functional derivatives with respect to the sources. We note that we set $j^\mu = 0$ *after* evaluating each of the n free field factors in eq. 8.15. The physical justification for this procedure is that we wish the free field propagators to be truly independent free field propagators and not depend on external sources, and not be convoluted together via the $Y^\mu$ field. The result will be a simpler, physically reasonable, perturbation theory in which each $\phi$ particle is truly a free field (independent of the other $\phi$ fields) that emits Y quanta and then absorbs all its emitted Y quanta. (An *alternative, different*, theory would allow a $\phi$ particle to emit a Y quantum that would then be absorbed by a different $\phi$ particle – thus $\phi$ particles could interact via the exchange of Y quanta. The exchange of Y quanta would make these seemingly free fields actually interacting fields – contrary to our starting asumption that the $\mathcal{L}_{F0i}$ terms each define a free field. This type of theory would also be a calculational nightmare since a $\phi$ particle emits an infinite superposition of Y quanta. Thus the lowest order diagram in $\phi^4$ perturbation would have an infinite number of terms. We therefore will not consider this possibility in the hope that Nature opted for our two-tier theory.)

We will now evaluate eq. 8.15 in stages since a number of novel features appear in its evaluation although the path integral evaluations themselves are conventional. We start by performing the Y integration. The Y field is a free abelian gauge field. We choose the same Coulomb gauge as in previous chapters $\nabla\cdot \mathbf{Y} = 0$ and $Y^0 = 0$. Consider

$$Z_Y(j_\mu) = \int D^4Y \exp\{ i\int d^4y[\mathcal{L}_C + j_\mu(y)Y^\mu(y)]\} \tag{8.16}$$

with $\mathcal{L}_C$ given by eq. 3.15. After some manipulations we find

$$Z_Y(j_\mu) = \exp\{ -i\int d^4y_1 d^4y_2\, j^i(y_1)D_{Fij}(y_1 - y_2)j^j(y_2)/2\} \tag{8.17}$$

with

$$D_{Fij}(y_1 - y_2) = \int d^4k\ e^{-ik\cdot(y_1 - y_2)}\ (\delta_{ij} - k_i k_j/\mathbf{k}^2)/[(2\pi)^4(k^2 + i\varepsilon)] \tag{8.18}$$

the spatial components of the massless vector boson Feynman propagator in the Coulomb gauge.

We now evaluate one of the factors in the exponentiated n free boson Lagrangian:

$$Z_{\phi_k}(J_k) = [\int D\phi_k \exp\{\ i\int d^4y\ [\mathcal{J}(-i\delta/\delta j^\nu(y), y)\mathcal{L}_{F0k} + J_k(y)\phi_k(X)]\}\ Z_Y(j_\mu)]\ |_{j_\mu = 0} \tag{8.19}$$

The free Klein-Gordon Lagrangian is:

$$\mathcal{L}_{F0k} = \tfrac{1}{2}\ [\ (\partial\phi_k/\partial X^\nu)^2 - m_k^{\ 2}\phi_k^{\ 2}] \tag{8.20}$$

As a result the classical action can be written as

$$S_{0k}[\phi_k] = \int d^4y\ \mathcal{J}(-i\delta/\delta j^\nu(y), y)\mathcal{L}_{F0k} \tag{8.21}$$

$$= -\tfrac{1}{2}\int d^4X_{op}\ \phi_k(X_{op})(\Box_X + m_k^{\ 2})\phi_k(X_{op}) \tag{8.22}$$

under the usual assumptions of good behavior at infinity and utilizing the Jacobian factor with

$$X_{op\mu}(y) = y_\mu + \ M_c^{-2}\ \delta/\delta j^\mu(y) \tag{8.23}$$

and with

$$\Box_X = (\partial/\partial X_{op}^{\ \nu})(\partial/\partial X_{op\nu}) \tag{8.24}$$

Inserting eq. 8.21 in eq. 8.19 we obtain:

$$Z_{\phi_k}(J_k) = [\int D\phi_k \exp\{i\int d^4X_{op}[-\tfrac{1}{2}\phi_k(X_{op})(\Box_X + m_k^{\ 2})\phi_k(X_{op}) +$$

$$+ J_k(y)\phi_k(y)/\mathcal{J}(-i\delta/\delta j^\nu(y), y)]\}\ Z_Y(j_\mu)]\ |_{j_\mu = 0} \tag{8.25}$$

The (Gaussian in $\phi_k$) path integral in eq. 8.25 can be performed (see below) using standard techniques to yield:

$$Z_{\phi_k}(J_k) = \Big[\exp\{-i\int d^4X_{op}(y_1)d^4X_{op}(y_2)[\,J_k(y_1)/\mathcal{J}(-i\delta/\delta j^\nu(y_1), y_1)]\cdot$$

$$\cdot\Delta_{Fk}{}^{TT}(y_1 - y_2)[\,J_k(y_2)/\mathcal{J}(-i\delta/\delta j^\nu(y_2), y_2)]/2\}\, Z_Y(j_\mu)\Big]\Big|_{j_\mu = 0}$$

$$= \exp\{-i\int d^4y_1 d^4y_2 J_k(y_1)\Delta_{Fk}{}^{TT}(y_1 - y_2)J_k(y_2)/2\} \tag{8.26}$$

after changing the integration variables in the exponential using the Jacobian with $\Delta_{Fk}{}^{TT}(y_1 - y_2)$ given by eqs. 6.51-52. The added index k serves to identify the propagator $\Delta_{Fk}{}^{TT}(y_1 - y_2)$ as having the mass $m_k$.

The derivation of eq. 8.25 uses functional techniques in a straightforward way. We can derive the form of $\Delta_{Fk}{}^{TT}(y_1 - y_2)$ best in momentum space. The operator $(\Box_X + m_k^2)^{-1}$ can be represented via

$$(\Box_X + m_k^2)^{-1}g(X_{op}(y_1))Z_Y(j_\mu) = (\Box_X + m_k^2)^{-1}\int d^4p\{e^{-ip\cdot X_{op}(y_1)}g(p)/(2\pi)^4\}Z_Y(j_\mu)$$

$$= -\int d^4p\{e^{-ip\cdot X_{op}(y_1)}g(p)/[(2\pi)^4(p^2 - m_k^2 + i\varepsilon)]\}Z_Y(j_\mu)$$

$$= -\int d^4X_{op}(y_2)\Delta_{Fkop}{}^{TT}(X_{op}(y_1) - X_{op}(y_2))g(X_{op}(y_2))Z_Y(j_\mu)$$

$$= -\int d^4X_{op}(y_2)\Delta_{Fkop}{}^{TT}(X_{op}(y_1) - X_{op}(y_2))Z_Y(j_\mu)g(X(y_2))$$

We now calculate

$$\Delta_{Fkop}{}^{TT}(X_{op}(y_1) - X_{op}(y_2))Z_Y(j_\mu) = \int d^4p\ \{e^{-ip\cdot(y_1 - y_2)}/[(2\pi)^4(p^2 - m_k^2 + i\varepsilon)]\}\cdot$$

$$\cdot\ \exp(-ip^\nu[\delta/\delta j^\nu(y_1) - \delta/\delta j^\nu(y_2)]/M_c^2)Z_Y(j_\mu)$$

$$= Z_Y(j_\mu)\ \int d^4p\ \{e^{-ip\cdot(y_1 - y_2)}/[(2\pi)^4\ (p^2 - m_k^2 + i\varepsilon)]\}\cdot$$

$$\cdot \exp(-ip^ip^j[D_{Fij}(y_1 - y_2) + D_{Fij}(y_2 - y_1)]/(2M_c^4)) \tag{8.27}$$

where $D_{ij}(y_1 - y_2)$ is given by eq. 8.18. The last factor in eq. 8.27 is the Gaussian damping factor that appears in two-tier propagators as repeatedly seen earlier – for example in eq. 6.53 and 6.54:

$$R(\mathbf{p}, y_1 - y_2) = \exp(-ip^ip^j[D_{Fij}(y_1 - y_2) + D_{Fij}(y_2 - y_1)]/(2M_c^4)) \tag{8.28}$$

Eq. 8.26 then follows with the two-tier scalar Klein-Gordon Feynman propagator.

Therefore the n scalar Klein-Gordon particle path integral of eq. 8.15 simplifies to

$$Z(J) = N\, \{\exp\{\, i\textstyle\int d^4y\, \mathcal{J}(-i\delta/\delta j^\nu(y), y)\, \mathcal{L}_{Fint}(-i\delta/\delta J_1(y), \ldots, -i\delta/\delta J_n(y))\}\cdot$$

$$\cdot\exp\{-\sum_k [\, i\textstyle\int d^4y_1 d^4y_2 J_k(y_1)\Delta_{Fk}{}^{TT}(y_1 - y_2)J_k(y_2)/2]\}\; Z_Y(j_\mu)\,\}\Big|_{j_\mu = 0} \tag{8.29}$$

The only dependence on functional derivatives with respect to $j^\mu$ in eq. 8.29 is in the Jacobian of the interaction Lagrangian term. It is readily seen that the Jacobian effectively reduces to one if there is no external physical source for Y particles (as we have assumed.) Consider

$$\begin{aligned}\mathcal{J}(-i\delta/\delta j^\nu(y),y)Z_Y(j_\mu) = \{\varepsilon^{ijk}&(\delta_{1i} + M_C^{-2}(\partial/\partial y^i)\delta/\delta j^1(y))\cdot\\ &\cdot(\delta_{2j} + M_C^{-2}(\partial/\partial y^j)\,\delta/\delta j^2(y))\cdot\\ &\cdot(\delta_{3k} + M_C^{-2}(\partial/\partial y^k)\,\delta/\delta j^3(y))Z_Y(j_\mu)\}\end{aligned} \tag{8.30a}$$

Since the interaction Hamiltonian – including the q-number Jacobian – is normal-ordered (eq. 6.8) the functional derivatives in eq. 8.30a can only apply to $Z_Y(j_\mu)$. Therefore

$$\mathcal{J}(-i\delta/\delta j^\nu(y),y)Z_Y(j_\mu) = Z_Y(j_\mu)\varepsilon^{ijk}\big(\delta_{1i} + M_C^{-2}(\partial/\partial y^i)(-i/2)[\textstyle\int d^4y_2 D_{F1a}(y - y_2)j^a(y_2) +$$

$$+ \textstyle\int d^4y_1\, j^a(y_1)D_{Fa1}(y_1 - y)]\big)\cdot\big(\delta_{2j} + M_C^{-2}(\partial/\partial y^j)\,(-i/2)[\textstyle\int d^4y_2\, D_{F2a}(y - y_2)j^a(y_2) +$$

$$+ \int d^4y_1\, j^a(y_1)D_{Fa2}(y_1 - y)]\Big)\cdot\Big(\delta_{3k} + M_C^{-2}(\partial/\partial y^k)\,(-i/2)[\int d^4y_2\, D_{F3a}(y - y_2)j^a(y_2) +$$

$$+ \int d^4y_1\, j^a(y_1)D_{Fa3}(y_1 - y)]\Big) \tag{8.30a}$$

After setting $j_\mu = 0$ *for the case of no incoming or outgoing Y quanta* we see

$$\mathcal{J}(-i\delta/\delta j^\nu(y),y) \equiv 1 \tag{8.30b}$$

This result is consistent with our physical notion that the integration, in itself, does not generate dynamical effects. *Otherwise* a c-number interaction Lagrangian term would generate a non-trivial interacting quantum field theory effects – contrary to our physical expectations. Eq. 8.30b eliminates this physically unreasonable possibility.

Thus eq. 8.29 becomes

$$Z(J) = N \exp\{\, i\int d^4y\, \mathscr{L}_{Fint}(-i\delta/\delta J_1(y), \ldots, -i\delta/\delta J_n(y))\}\cdot$$

$$\cdot\exp\{-i\int d^4y_1 d^4y_2 \sum_k J_k(y_1)\Delta_{Fk}{}^{TT}(y_1 - y_2)J_k(y_2)/2]\} \tag{8.31}$$

which gives the same perturbation theory found earlier using canonical quantum field theory that is built on the U matrix expansion.

*Derivative Coupling Case*

Eq. 8.31 is based on the assumption that derivative couplings do not appear in the interaction Lagrangian. (There are no derivative couplings in the Standard Model but there are derivative couplings in quantum gravity so we must deal with this issue if we wish to create a unified theory.) If the interaction Lagrangian does contain derivatives of scalar fields with respect to the X variable then the interaction Lagrangian has a superficial dependence on the functional derivative with respect to $j^\mu$, which we can symbolize in the modified path integral expression:

$$Z(J) = N\, \{\exp\{\, i\int d^4y\, \mathscr{L}_{Fint}(\partial/\partial(y^\nu + M_c^{-2}\delta/\delta j_\nu(y)), -i\delta/\delta J_1(y), \ldots, -i\delta/\delta J_n(y))\}\cdot$$

$$\cdot\exp\{-\sum_k [\, i\int d^4y_1 d^4y_2 J_k(y_1)\Delta_{Fk}{}^{TT}(y_1 - y_2)J_k(y_2)/2]\}\; Z_Y(j_\mu)\,\}\,\big|_{j_\mu = 0} \tag{8.32}$$

obtained from eq. 8.31.

The $j^\mu$ dependence now appears only in the interaction Lagrangian. For good reason we will now show the $j^\mu$ dependence of the interaction Lagrangian in eq. 8.32 can be eliminated.

Our approach will be to separate the coordinate dependence in the propagator into two parts: the coordinate dependence in the Gaussian factor, and the coordinate dependence in the $e^{ip\cdot x}$ factor. We can then express the derivatives of fields in the interaction lagrangian as derivatives with respect to the coordinates in the $e^{ip\cdot x}$ factor appearing in integral representations of the two-tier propagator when a perturbative expansion of the path integral solution is made.

We begin by noting that

$$i\Delta_{Fk}{}^{TT}(y_1 - y_2) = <0|T(\phi(X(y_1)),\phi(X(y_2)))|0> \quad (6.51)$$

$$= i\int \frac{d^4p\, e^{-ip\cdot(y_1 - y_2)} R(\mathbf{p}, y_1 - y_2)}{(2\pi)^4 (p^2 - m_k{}^2 + i\varepsilon)} \quad (6.52)$$

with

$$R(\mathbf{p}, y_1 - y_2) = \exp[-p^i p^j \Delta_{Tij}(y_1 - y_2)/M_c{}^4] \quad (6.53)$$

(summations are over space indices only in the Y Coulomb gauge) and

$$\Delta_{Tij}(z) = \int d^3k\, e^{-ik\cdot z} (\delta_{ij} - k_i k_j/\mathbf{k}^2)/[(2\pi)^3 2\omega_k] \quad (6.54)$$

We now define *a more general* two-tier propagator by introducing a distinction between the spatial dependence of the gaussian and exponential terms:

$$i\Delta_{Fk}{}^{TT}(y_1 - y_2, z) = i\int \frac{d^4p\, e^{-ip\cdot(y_1 - y_2)} R(\mathbf{p}, z)}{(2\pi)^4 (p^2 - m_k{}^2 + i\varepsilon)} \quad (8.33)$$

We then note that

$$\partial i\Delta_{Fk}{}^{TT}(y_1 - y_2)/\partial X^\mu = \partial i\Delta_{Fk}{}^{TT}(y_1 - y_2, z)/\partial y_1{}^\mu |_{z=y_1 - y_2} \quad (8.34)$$

comparing eqs. 6.51 and 6.52 with eq. 8.33. As a result we can write eq. 8.32 symbolically as:

$$Z(J) = N\{\exp\{ i\int d^4y\, \mathscr{L}_{Fint}(\partial/\partial y^\nu, -i\delta/\delta J_1(y), \ldots, -i\delta/\delta J_n(y))\}\cdot$$

$$\cdot \exp\{-\sum_k [\, i\int d^4y_1 d^4y_2 J_k(y_1)\Delta_{Fk}{}^{TT}(y_1 - y_2, z)J_k(y_2)/2]\}\} \,|_{z=y_1-y_2} \tag{8.35}$$

Eq. 8.35 is interpreted as the following:

1. For a given process take appropriate functional derivatives of Z(J) with respect to $J_1(y), \ldots, J_n(y)$.

2. Then expand the exponential factors in a perturbation series applying any derivatives with respect to y in $\mathscr{L}_{Fint}$. Do not perform any of the $\int d^4y_1 d^4y_2$ integrals.

3. Then set $z = y_1 - y_2$ in each $\Delta_{Fk}{}^{TT}(y_1 - y_2, z)$ propagator.

4. Lastly perform all $\int d^4y_1 d^4y_2$ integrals.

Thus we thus achieve a path integral formulation that is very similar to the corresponding expression in conventional field theory – the only difference is the form of the free field propagators, which now contain a Gaussian factor.

*$\phi^4$ Theory Path Integral Formulation Example*

We will now consider the case of two-tier scalar field theory using the specific example of the $\phi^4$ Lagrangian of eqs. 5.20 – 5.24:

$$\mathscr{L}_{Fint}(X^\mu) = \tfrac{1}{4!}\, \chi_0\, \phi(X)^4 + \tfrac{1}{2}\,(m^2 - m_0{}^2)\phi(X)^2 \tag{8.36}$$

with

$$\mathscr{L}_{F0} = \tfrac{1}{2}\,[\, (\partial\phi/\partial X^\nu)^2 - m^2\phi^2] \tag{8.37}$$

In this theory eq. 8.31 can be written as:

$$Z(J) = N \exp\{i\int d^4y\, \mathscr{L}_{Fint}(-i\delta/\delta J(y))\}\, \exp\{-i\int d^4y_1 d^4y_2 J(y_1)\Delta_{Fk}{}^{TT}(y_1 - y_2)J(y_2)/2\} \tag{8.38}$$

The perturbation theory generated from eq. 8.38 is the same as conventional perturbation theory except for the differing free field propagators.

## Two-Tier Yang-Mills Gauge Fields

Two-tier Yang-Mills gauge field theories have many similarities to conventional Yang-Mills theories.[39] We assume the reader in familiar with internal symmetries and the conventional Yang-Mills formulation.

*General Rule:* All gauge fields and derivatives of gauge fields, as well as group properties and other features such as the Fadeev-Popov method, are expressed solely in terms of the X coordinate system (which in turn is a function of the y coordinates).

We note all matter fields are assumed to be functions of X coordinates only as done in previous chapters. The general rule is implemented by defining:

1. The covariant derivative of any matter field $\Psi(X)$ by

$$D^\mu\Psi(X) = [\partial/\partial X_\mu + igA^\mu(X)]\Psi(X) \tag{8.39}$$

with $A^\mu(X)$ being an element of a Lie algebra:

$$A^\mu(X) = A_a{}^\mu(X)L_a \tag{8.40}$$

where $L_a$ is a generator of a Lie algebra (a and b are internal symmetry indexes) with commutation relations:

$$[L_a, L_b] = ic_{ab}{}^c L_c \tag{8.41}$$

with $c_{ab}{}^c$ being real numbers called the *structure constants* of the Lie algebra.

2. The field strengths are defined as the commutator of covariant derivatives:

$$F^{\mu\nu} = [D^\mu, D^\nu] = \partial A^\nu(X)/\partial X_\mu - \partial A^\mu(X)/\partial X_\nu + ig[A^\mu(X), A^\nu(X)] \tag{8.42}$$

3. The Lagrangian for a Yang-Mills gauge field interacting with a matter field $\psi(X)$ has the form:

$$\mathscr{L}_{YM} = [\mathscr{L}_F^{YM}(X^\mu) + \mathscr{L}_F^{Matter}(X^\mu)]\, J + \mathscr{L}_C(X^\mu(y), \partial X^\mu(y)/\partial y^\nu) \tag{8.43}$$

where J is the Jacobian and

---

[39] C. N. Yang and R. L. Mills, Phys. Rev. **96**, 191 (1954).

$$\mathscr{L}_F^{YM}(X^\mu) = -\tfrac{1}{4}\, F_a^{\mu\nu}\, F_{a\mu\nu} \tag{8.44}$$

with $F^{\mu\nu} = F_a^{\mu\nu} L_a$ and

$$\mathscr{L}_F^{Matter}(X^\mu) = \mathscr{L}_F^{Matter}(\psi(X), D^\mu\psi(X)) \tag{8.45}$$

and $\mathscr{L}_C$ is defined as previously and specifies the Y field evolution.

The generalization to multiple gauge fields interacting with multiple matter fields is direct. The overall form of the Standard Model Lagrangian is:

$$L = \int d^4y\, \{J[\mathscr{L}_F^{Matter}(X^\mu) + \mathscr{L}_F^{GaugeFields}(X^\mu) + \mathscr{L}_F^{Higgs}(X^\mu)] + \\ + \mathscr{L}_C(X^\mu(y), \partial X^\mu(y)/\partial y^\nu)\} \tag{8.46}$$

where J is the Jacobian for the transformation from y to X coordinate integration. Eq. 8.46 can be rewritten as

$$L = \int d^4X[\mathscr{L}_F^{Matter}(X^\mu) + \mathscr{L}_F^{GaugeFields}(X^\mu) + \mathscr{L}_F^{Higgs}(X^\mu)] + \\ + \int d^4y \mathscr{L}_C(X^\mu(y), \partial X^\mu(y)/\partial y^\nu) \tag{8.47}$$

It is clear from eq. 8.47 that the conventional Standard Model equations of motion and canonical quantization procedure emerge in the two-tier formulation if all expressions are written as functions of the X coordinates as shown previously in our discussions of separable Lagrangians. The second quantization of X as a function of the y coordinates leads to the gaussian factor in all free particle propagators (except the Y propagator).

Thus the gauge field and matter field parts of the Lagrangian, the gauge field transformation laws and other related operations solely depend on the X coordinate system as stated in the general rule above. The quantization and field equations are the same as the conventional case – except that they are specified solely in terms of the X coordinate system.

### *Path Integral Formulation and Fadeev-Popov Method*

We now turn to the two-tier path integral formulation of Yang-Mills gauge theories and in particular to the two-tier version of the Fadeev-Popov method. The two-tier path integral for a gauge field can be written symbolically as:

$$Z(J^\mu) = N \int DADY\Delta_{FP}(A)\delta(F(A)) \exp\{i\int d^4y\, [\mathscr{L} +$$

$$+ \ j_\mu(y)Y^\mu(y)+J^\mu(y)A_\mu(X)]\} \,|_{j_\mu = 0} \tag{8.48}$$

where $\delta(F(A))$ specifies the gauge and $\Delta_{FP}(A)$ is the Fadeev-Popov determinant. The Lagrangian is

$$\mathscr{L} = J\mathscr{L}_F^{YM}(X^\mu) \ + \mathscr{L}_C(X^\mu(y), \partial X^\mu(y)/\partial y^\nu) \tag{8.49}$$

with $\mathscr{L}_F^{YM}$ specified by eq. 8.44 and J the Jacobian for the transformation from X coordinates to y coordinates. The Fadeev-Popov determinant may be calculated in the standard way. First we note that the delta function fixing the gauge can be written as a delta function in the gauge times a determinant:

$$\delta(F(A^\omega)) = \delta(\omega - \omega_0) \,|\det \delta F(A_\mu{}^\omega(X))/\delta\omega(X)\,|^{-1} \Big|_{F(A)=0} \tag{8.50}$$

where $\omega_0$ is a reference gauge, where

$$A_\mu{}^\omega(X) = A_\mu(X) + \partial\omega(X)/\partial X^\mu \tag{8.51}$$

and where

$$\Delta_{FP}(A) = \ |\det \delta F(A_\mu{}^\omega(X))/\delta\omega(X)\,| \ \Big|_{F(A)=0} \tag{8.52}$$

We will choose the Lorentz gauge to evaluate the Fadeev-Popov determinant:

$$F_a(A) = \partial A_a{}^\mu(X)/\partial X^\mu = 0 \tag{8.53}$$

Under an infinitesimal gauge transformation:

$$A_{a\mu}{}^\omega(X) = A_{a\mu}(X) + g^{-1}\partial\omega_a/\partial X^\mu + c_{ab}{}^c\,\omega_b(X)A_c{}^\mu(X) \tag{8.54}$$

we find

$$F_a(A_\mu{}^\omega(X)) = \partial(A_{a\mu}(X) + g^{-1}\partial\,\omega_a(X)/\partial X^\mu + c_{ab}{}^c\,\omega_b(X)A_c{}^\mu(X))/\partial X^\mu$$

$$= g^{-1}\,\Box_X\,\omega_a(X) + c_{ab}{}^c\,\partial\,\omega_b(X)/\partial X^\mu\,A_c{}^\mu(X) \tag{8.55}$$

Thus

$$\delta F_a(A_\mu{}^\omega(X))/\delta\omega_b(X) = g^{-1}\,\delta_{ab}\Box_X + c_{ab}{}^c A_c^\mu(X)\partial/\partial X^\mu \tag{8.56}$$

and

$$\Delta_{FP}(A) = \left|\det\left(g^{-1}\,\delta_{ab}\Box_X + c_{ab}{}^c A_c^\mu(X)\partial/\partial X^\mu\right)\right|\,\Big|_{F(A)=0} \tag{8.57}$$

where | ... | represent absolute value.

We note the two-tier Fadeev-Popov determinant is solely a function of the X coordinates. Thus we can follow the standard procedure and rewrite the determinant as a path integral over anti-commuting c-number fields with a ghost Lagrangian that is solely a function of X – just like the gauge particles and other particles in two-tier theories:

$$\Delta_{FP}(A) = \int D\chi^* D\chi \, \exp[\, i\int d^4X\, \mathscr{L}^{ghost}(X^\mu)] \tag{8.58}$$

where

$$\mathscr{L}^{ghost}(X^\mu) = \chi_a{}^*(X)[\delta_{ab}\Box_X + g\, c_{ab}{}^c A_c^\mu(X)\partial/\partial X^\mu]\chi_b(X) \tag{8.59}$$

Thus the complete path integral is

$$Z(J^\mu) = N\int DAD\chi^* D\chi DY\delta(F(A))\,\exp\{i\int d^4y\,[\mathscr{L} +$$

$$+\, j_\mu(y)Y^\mu(y) + J^\mu(y)A_\mu(X)]\}\,|_{j_\mu = 0} \tag{8.60}$$

where $\delta(F(A))$ specifies the gauge and

$$\mathscr{L} = J\mathscr{L}_F{}^{YM}(X^\mu) + J\mathscr{L}^{ghost}(X^\mu) + \mathscr{L}_C(X^\mu(y), \partial X^\mu(y)/\partial y^\nu) \tag{8.61}$$

with $\mathscr{L}_F^{YM}$ specified by eq. 8.44.

At this point it is obvious that we can follow almost identical steps as we did in the scalar particle case starting from eq. 8.9 and obtain an expression similar to eq. 8.38. The result is a perturbation theory for the Yang-Mills gauge field that is identical to the usual theory except that the free propagators for the gauge fields, *and the ghost fields*, acquire the gaussian factor R(p,z) as stated earlier in the discussions of eqs. 6.52, 6.88, 6.95, 6.102, and 7.120-123.

## Two-Tier Massive Vector Fields

Massive vector fields have been a problem for perturbation theory due to the $k_\mu k_\nu/m^2$ term appearing in the free field propagator. This term makes conventional interacting quantum field theories of massive vector bosons non-renormalizable. The Higgs mechanism is used in Electroweak theory to evade the non-renormalizability of massive gauge fields. It gives mass to the vector bosons mediating the weak force while maintaining the renormalizability of the theory. It also implements symmetry breaking.

*In Two-Tier quantum field theory a massive vector boson does not create renormalization issues. For example the two-tier version of the Weinberg-Salam model of the electromagnetic and weak forces (with massive vector bosons) is finite! Thus the need for spontaneous symmetry breaking to give mass to vector bosons in Electroweak theory is not present. Higgs particles are not needed (although they may exist. Their existence is an experimental question – not a theoretical necessity for a two-tier Electroweak theory with massive vector bosons.)*

### *Massive Vector Boson Propagator*

The massive free vector particle propagator in <u>*conventional*</u> quantum field theory has the representation:

$$i\Delta_{FV}(y_1 - y_2)_{\mu\nu} = -i \int \frac{d^4k\, e^{-ik\cdot(y_1 - y_2)}\,(g_{\mu\nu} - k_\mu k_\nu/m^2)}{(2\pi)^4\,(k^2 - m^2 + i\varepsilon)} \tag{8.62}$$

A two-tier massive vector boson theory can be constructed in a straightforward way from the Lagrangian:

$$\mathscr{L} = J[-\tfrac{1}{4} F_V^{\mu\nu}(X(y))F_{V\mu\nu}(X(y)) - \tfrac{1}{2} m^2 V^\mu V_\mu] + \mathscr{L}_C(X^\mu(y), \partial X^\mu(y)/\partial y^\nu, y) \tag{8.63}$$

with J the Jacobian and with

$$F_{V\mu\nu}(X(y)) = \partial V_\mu(X(y))/\partial X^\nu - \partial V_\nu(X(y))\,/\partial X^\mu \tag{8.64}$$

and the usual two-tier Y Lagrangian terms

$$\mathscr{L}_C(X^\mu(y), \partial X^\mu(y)/\partial y^\nu, y) = -\tfrac{1}{4} F_Y^{\mu\nu}F_{Y\mu\nu} \tag{7.4}$$

$$F_{Y\mu\nu} = (\partial Y_\mu/\partial y^\nu - \partial Y_\nu/\partial y^\mu) \tag{7.5}$$

Following steps similar to the previously considered scalar particle, fermion and massless vector particle (photon) cases we find the two-tier massive vector boson Feynman propagator to be:

$$i\Delta_{FV}{}^{TT}(y_1 - y_2)_{\mu\nu} = -i \int \frac{d^4k\, e^{-ik\cdot(y_1 - y_2)}\, (g_{\mu\nu} - k_\mu k_\nu/m^2)\, R(\mathbf{k}, y_1 - y_2)}{(2\pi)^4\, (k^2 - m^2 + i\varepsilon)} \tag{8.65}$$

*No Need for Higgs Mechanism in Electroweak Theory*

The leading coordinate space dependence at high energy (short distance) of the fourier transform of $\Delta_{FV}{}^{TT}$ is the same as $\Delta_F{}^{TT}$ since it comes from the $g_{\mu\nu}$ term in $\Delta_F{}^{TT}$

$$\Delta_{FV}{}^{TT}(y_1 - y_2)_{\mu\nu} \sim g_{\mu\nu}\, (y_1 - y_2)^2 \tag{8.66}$$

which in momentum space is equivalent to

$$\Delta_{FV}{}^{TT}(p)_{\mu\nu} \sim g_{\mu\nu}\, p^{-6} \tag{8.67}$$

The $k_\mu k_\nu/m^2$ term appearing in the free vector field two-tier propagator actually is of higher order in the large momentum limit (short distance):

$$\int \frac{d^4k\, e^{-ik\cdot(y_1 - y_2)} k_\mu k_\nu/m^2)\, R(\mathbf{k}, y_1 - y_2)}{(2\pi)^4\, (k^2 - m^2 + i\varepsilon)} \sim (y_1 - y_2)^4 \tag{8.68}$$

which corresponds to momentum space behavior of

$$p^{-8} \tag{8.69}$$

*Two-Tier propagators such as the propagator in eq. 8.65 have the feature that the gaussian factor "inverts" the high-energy behavior of the terms in the numerator of the integrand: terms with higher powers of momentum are less significant at short distances. The term with the lowest power of momentum generates the leading behavior at high energy (short distances).*

Thus two-tier massive vector boson theories are ultra-violet convergent and do not constitute a problem as they do in conventional quantum field theory. *Therefore there is no need for the Higgs mechanism in the Electroweak sector of the Standard Model in order to obtain a renormalizable theory. Ordinary massive vector bosons can be used and the resulting two-tier theory is finite!*

## General Short Distance (High Momentum) Behavior of Two-Tier Propagator

*The higher the power of the momentum in the numerator of the integrand of a two-tier propagator, the more convergent the large momentum behavior of the fourier transform of the two-tier propagator.*

The short distance behavior of a term with n factors of momentum in a two-tier propagator has the leading short distance coordinate space behavior:

$$\int \frac{d^4k\, e^{-ik\cdot(y_1 - y_2)}\, k_{\mu_1} k_{\mu_2} \ldots k_{\mu_n}\, R(\mathbf{k}, y_1 - y_2)}{(2\pi)^4\, (k^2 - m^2 + i\varepsilon)} \sim (y_1 - y_2)^{2+n} \tag{8.70}$$

which corresponds to the high energy behavior:

$$p^{-6-n} \tag{8.71}$$

*In contrast to conventional quantum field theory the more powers of momentum in the numerator of a two-tier propagator, the better the short distance behavior!*

## Higgs Particles

A previous section shows that the Higgs Mechanism is not needed in the two-tier Electroweak sector of the Standard Model. Massive vector bosons such as **W**'s would not make the Electroweak theory non-renormalizable. *As we have seen a Two-tier Electroweak theory with massive vector bosons is finite.*

Nevertheless we would like to point out a two-tier version of the Higgs particle sector of the Standard Model can be defined that largely parallels the conventional treatment of the Higgs sector. Higgs particles continue to be of interest since they may play a role in the origin of particle masses and symmetry breaking.

The two-tier scalar Higgs field Lagrangian terms (plus $\mathscr{L}_C$) can be written:

$$\mathscr{L}_{\text{Higgs}} = J[D_\mu\phi^\dagger D^\mu\phi - V(\phi(X))] + \mathscr{L}_C \tag{8.72}$$

$$V(\phi(X)) = m_0^2\phi^\dagger(X(y))\phi(X(y)) + \lambda[\phi^\dagger(X(y))\phi(X(y))]^2 + \\ + G_c[\bar{L}(X)\phi(X)R(X) + \bar{R}(X)\phi^\dagger(X)R(X)] \qquad (8.73)$$

with the covariant derivative defined with the usual $B_\mu$ and $W_\mu^a$ gauge fields of the Standard Model, and L representing a left-handed fermion isodoublet, and R representing a right-handed fermion isosinglet. We note all items in eqs. 8.72 and 8.73 are written solely as functions of the X coordinates with the exception of $\mathscr{L}_C$, which is the Lagrangian term for the Y field.

The conventional effective potential method can be followed to implement the Higgs mechanism. In particular we may write

$$\phi(X(y)) = <\phi> + \eta(X) \qquad (8.74)$$

with $<\phi>$ the vacuum expectation value, which is a constant by translational invariance. Then vector bosons can acquire mass via the Higgs mechanism. The quantum part of the Higgs field has a two-tier propagator with $p^{-6}$ behavior for large momentum. Thus all sectors of the Standard Model wind up with two-tier propagators and all perturbative calculations are finite.

## General Form of the Two-Tier Standard Model Path Integral

The general form of the path integral for the two-tier version of the Standard Model is:

$$Z(J) = N\,\{\exp\{\,i\int d^4y\,\mathscr{L}_{Fint}(\partial/\partial y^\nu, -i\delta/\delta J_1(y), \ldots, -i\delta/\delta J_n(y))\}\cdot$$

$$\cdot \exp\{-\sum_k [\,i\int d^4y_1 d^4y_2 J_k(y_1)\Delta_{Fk}^{TT}(y_1 - y_2, z)J_k(y_2)/2]\}\}\,|_{z=y1-y2} \qquad (8.75)$$

where the sum over k is a sum over all matter fields, gauge fields, Higgs fields, and ghost fields. The index k, and indices on the functional derivatives, represent all space-time indices and internal symmetry indices that are relevant for each particle. We assume the total number of particle and ghost fields is n. Notice all dependence on the Y field has been "integrated away."

Also any derivatives with respect to y appearing in the interaction Lagrangian are applied to two-tier propagators using the method represented by eqs. 8.33 and 8.34. This procedure results in the same momentum polynomials for Feynman diagrams in a two-tier version as appear in the corresponding conventional theory.

Thus we have the same algebraic structure (both in the momentum polynomials and internal symmetries) in the two-tier version of a conventional theory.

We note that the large distance behavior of the two-tier theory is the same as the Standard Model in all respects including gauge symmetries. Deviations from the conventional Standard Model results only appear at extremely high energies of the order of $M_c$.

## Renormalization - Finite

Since all particle (and ghost) propagators are two-tier propagators in the two-tier Standard Model, the two-tier Standard Model yields finite results to all orders in perturbation theory. We note the large momentum behavior of the various types of particle propagators in the two-tier Standard Model is:

<u>Fermion Propagators</u>

$$p^{-7}$$

<u>Vector Boson (Gauge Field) Propagators</u>

$$p^{-6}$$

<u>Ghost Propagators</u>

$$p^{-6}$$

<u>Higgs Particle Propagators</u>

$$p^{-6}$$

Thus all Feynman diagrams are highly ultra-violet convergent and the two-tier Standard Model is finite. This result is independent of the details of the internal symmetries, particle spectrum, and particle masses.

## Unitarity

The two-tier Standard Model *superficially* appears to have a unitarity problem due to the non-hermitean nature of its hamiltonian. The lack of hermiticity is due entirely to the appearance of $iY^\mu$ in the $X^\mu$ field coordinates.

Thus the two-tier Standard Model interaction hamiltonian is not hermitean:

$$H_{Fint} = \int d^3y' \mathscr{H}_{Fint}( y' + iY(y')/M_c^2) \tag{8.76}$$

and

$$H_{Fint}{}^{\dagger} = \int d^3y' \mathscr{H}_{Fint}( y' - iY(y')/M_c{}^2) \neq H_{Fint} \tag{8.77}$$

The relation between $H_{Fint}$ and its hermitean conjugate is

$$H_{Fint} = V\, H_{Fint}{}^{\dagger}\, V \tag{8.78}$$

where $V^2 = I$ is the metric operator defined in eqs. 5.16 – 5.18. Eq. 8.78 implies that the two-tier Standard Model S matrix is not unitary. The two-tier Standard Model S matrix is pseudo-unitary:

$$S^{-1} = V\, S^{\dagger}\, V \tag{8.79}$$

Therefore

$$S^{\dagger} VS = V \tag{8.80}$$

*Two-Tier Standard Model S matrix satisfies unitarity between physical asymptotic states* – states consisting of only physical particles: leptons, quarks, photons, W and Z particles, gluons, and Higgs particles (if they exist). Put another way: physical states can consist of any set of particles in the two-tier Standard Model except ghosts and Y particles. The proof is identical to eqs. 6.46 – 6.48.

## Anomalies

The axial anomaly (Adler-Bell-Jackiw anomaly) follows from the linear divergence of a fermion triangle graph (Fig. 8.1) in the conventional Standard Model. All higher order terms are divergence-free. These terms do not contribute to the axial anomaly. Thus the axial anomaly can properly be regarded as an artifact of the regularization of the divergence of the fermion triangle diagram.

In two-tier theory the axial anomaly does not appear to be present. Fermion triangle diagrams in two-tier quantum field theories are finite. Thus the source of the anomaly in conventional theories is absent in two-tier theories.

A massless Dirac field theory is formally invariant under a chiral transformation implying a conserved axial-vector current. The two-tier axial-vector current is

$$j_5{}^{\mu}(X(y)) = \bar{\psi}(X(y))\gamma^{\mu}\gamma_5\psi(X(y)) \tag{8.84}$$

with formal conservation law:

$$\partial j_5{}^{\mu}(X(y))/\partial X^{\mu} = 2m\; j_5(X(y)) = 2m\; \bar{\psi}(X(y))\gamma_5\psi(X(y)) \tag{8.85}$$

Eq. 8.85 implies

$$\partial j_5^{\mu}(X(y))/\partial X^{\mu} = 0 \tag{8.86}$$

in the limit $m \rightarrow 0$. The question we now address is whether eq. 8.86 holds in two-tier perturbation theory – perhaps in the same form as the conventional axial anomaly:

$$\partial j_5^{\mu}(X(y))/\partial X^{\mu} = 2m\, j_5(X(y)) + \alpha_0(4\pi)^{-1}\varepsilon^{\mu\nu\alpha\beta}F_{\alpha\beta}F_{\mu\nu} \quad ? \tag{8.87}$$

where $\alpha_0$ is the unrenormalized fine structure constant.

The simplest manifestation of the axial anomaly in conventional field theory is the fermion triangle diagram, which we will now examine in two-tier quantum field theory. As stated earlier, the two-tier triangle diagram is finite and zero unlike the conventional quantum field theory result. *Thus the axial anomaly does not appear to exist in two-tier quantum field theory. The axial anomaly is a result of the divergence of the triangle diagram in conventional quantum field theory.*

The absence of the anomaly reflects the absence of divergences in two-tier quantum field theory, which preserves chiral invariance. Unlike Pauli-Villars regularization, for example, the finiteness of two-tier theory follows from the Gaussian factors. Unlike the dimensional regularization approach (where there is no equivalent to $\gamma_5$), two-tier theory can use the normal $\gamma_5$ matrix.

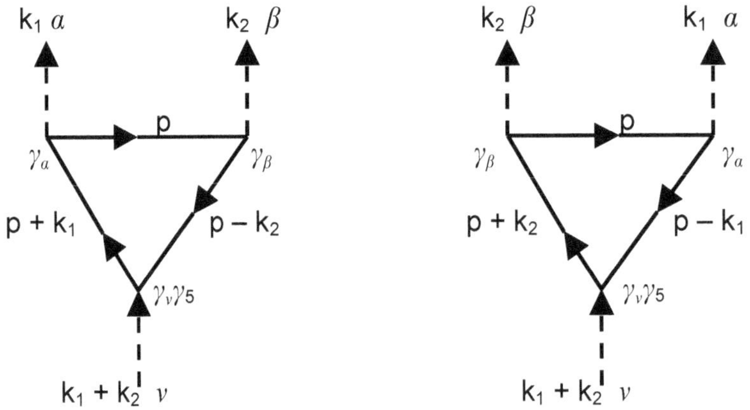


Figure 8.1. The V-V-A triangle diagrams.

The expression for the two-tier triangle diagrams is:

$$T_{\alpha\beta\nu}(k_1, k_2) = S_{\alpha\beta\nu}(k_1, k_2) + S_{\beta\alpha\nu}(k_2, k_1) \tag{8.88}$$

where

$$S_{\alpha\beta\nu}(k_1, k_2)\delta^4(k_1 + k_2 - q) = -iN\int d^4y_1 d^4y_2 d^4y_3\ e^{ik_1\cdot y_1 + ik_2\cdot y_2 - iq\cdot y_3} \cdot$$
$$\cdot\ Tr\{S_F^{\ TT}(y_1 - y_3)\gamma_\alpha S_F^{\ TT}(y_2 - y_1)\gamma_\beta S_F^{\ TT}(y_3 - y_2)\gamma_\nu\gamma_5\} / (2\pi)^4 \tag{8.89}$$

where N is a constant, and where $S_F^{\ TT}(z)$ is specified by eq. 7.45. We now define the fourier transform:

$$S_F^{\ TT}(z) = -i\int d^4p\ e^{-ip\cdot z}\ \mathscr{S}^{TT}(p) / (2\pi)^4 \tag{8.90}$$

where $\mathscr{S}^{TT}(p)$ defined by eq. 7.93. We then substitute the fourier transform in eq. 8.89 and perform the coordinate integrations to obtain:

$$S_{\alpha\beta\nu}(k_1, k_2) = N\int d^4p\ Tr\{\mathscr{S}^{TT}(p + k_1)\gamma_\alpha \mathscr{S}^{TT}(p)\gamma_\beta \mathscr{S}^{TT}(p - k_2)\gamma_\nu\gamma_5\} / (2\pi)^4$$

We note that

$$k_1^{\ \alpha} T_{\alpha\beta\nu}(k_1, k_2) \neq 0 \tag{8.91}$$

$$k_2^{\ \beta} T_{\alpha\beta\nu}(k_1, k_2) \neq 0 \tag{8.92}$$

$$(k_1 + k_2)^\nu T_{\alpha\beta\nu}(k_1, k_2) \neq 0 \tag{8.93}$$

in two-tier quantum field theory because the conservation laws are expressed with respect to the X coordinates – not the y coordinates. Thus since $k_1^{\ \alpha}$ corresponds $\partial/\partial y_\alpha$, and not $\partial/\partial X_\alpha$ there is no reason for eqs. 8.91-93 to be zero. However at "large distances" relative to $M_c^{-1}$ we see

$$k_1^{\ \alpha} T_{\alpha\beta\nu}(k_1, k_2) \cong 0 \tag{8.94}$$

$$k_2^{\ \beta} T_{\alpha\beta\nu}(k_1, k_2) \cong 0 \tag{8.95}$$

$$(k_1 + k_2)^\nu T_{\alpha\beta\nu}(k_1, k_2) \cong 0 \tag{8.96}$$

to very good approximation since the gaussian damping factor in the fermion propagators is approximately unity and thus the two-tier expression becomes essentially the same as the conventional field theory expression.

On the other hand at very short distances the anomaly appears to be absent since two-tier theory is very well behaved at high energy with

$$\mathscr{S}^{TT}(p) \backsim \gamma^0 M_c^{\,6}\, p^{-7} + \mathcal{O}(p^{-9}) \qquad (7.102)$$

As a result we see

$$k_1^{\,\alpha} T_{\alpha\beta\nu}(k_1, k_2) \sim p^{4-21} \sim p^{-17} \qquad (8.97)$$

as $p \to \infty$ is highly convergent. Thus there is no high energy divergence unlike conventional field theory where the integral is linearly divergent. And so no anomaly is generated.

## Asymptotic Freedom and Quark Confinement

Two-tier quantum field theory is totally consistent with the Standard Model at currently accessible energies. Thus if the Standard Model Quantum Chromodynamics (QCD) sector is asymptotically free, and also gives quark confinement at large distance (compared to $M_c^{-1}$), then similar statements would also be true in the two-tier version of QCD.

We assume $M_c$ is extremely large – much beyond current energies – and possibly of the order of the Planck mass. Fig. 8.2 depicts the various regions in two-tier QCD assuming a very large $M_c$.

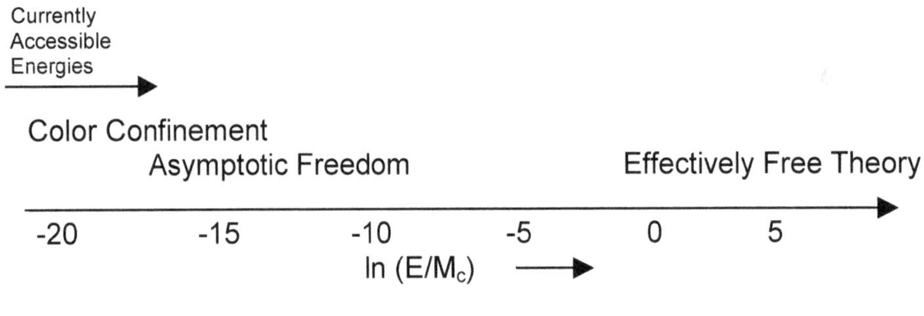


Figure 8.2. A depiction of the two-tier QCD regions as a function of the logarithm of the energy assuming $M_c$ is of the order of the Planck mass.

## Two-Tier Coordinates

We have developed a physical picture of two-tier coordinate systems in which we define two sets of related coordinates. This picture views ordinary real 4-dimensional space as a low energy approximation to a complex 4-dimensional space that only becomes apparent at ultra-high energies with the scale set by $M_c$. The imaginary part of the coordinates is based on the excitations of a quantum field $Y^\mu$. The real part of the coordinates is the familiar 4-dimensional c-number coordinate system.

Another way to view the two-tier X and y coordinate systems is to view them as defining a four-dimensional hyperplane in a six (or eight) dimensional real space-time. See Blaha (2004) for a detailed discussion.

It would also be interesting to consider an extension of the theory to the case where both the real and imaginary parts of the coordinate system are quantum fields. In this case the c-number coordinates of daily experience would emerge as a "condensation", or spontaneous symmetry breaking, phenomena.

One could also generalize the Y quantum field to a non-abelian gauge field, which, properly handled, could be the origin of internal symmetries such as QCD SU(3) using a Kaluza-Klein approach within a general relativistic *vierbein* framework. This approach is an alternative to the compactification of dimensions that is a cornerstone of SuperString theories.

# 9. Two-Tier Quantum Gravity: Finite!

## Introduction

There are numerous excellent books and monographs on classical gravity and a large literature on quantum gravity.[40] Therefore our discussion will assume the reader is familiar with classical General Relativity and aware of attempts to create quantum theories of gravity.

We will begin by establishing the general form of two-tier classical General Relativity and then proceed to define a quantization procedure. We will work in Minkowski space with three space and one time dimension. The flat-space metric $\eta_{\alpha\beta}$ is defined as diagonal with $\eta_{00} = 1$ and $\eta_{ij} = -\delta_{ij}$ for i, j = 1, 2, 3.

## Two-Tier General Relativity

In developing Quantum Gravity we will make the same ansatz that we have made throughout our development of two-tier quantum field theories: all field expressions are functions of the X coordinate field system, which in turn are functions of the "ordinary" y space-time coordinate system. Two-Tier Theory of Quantum Gravity is invariant under special relativistic transformations. The dynamical field equations, which are strictly functions of the X coordinates, are covariant under general relativistic transformations. The rationale for these assumptions is described in detail in chapter 10.

---

[40] H. Weyl, *Space, Time, Matter* (Dover, New York, 1950); L. D. Landau and E. M. Lifshitz, *The Classical Theory of Fields*, (Addison-Wesley, New York, 1962); S. Weinberg, *Gravitation and Cosmology*, (John Wiley & Sons, New York, 1972); C. W. Misner, K. S. Thorne and J. A. Wheeler, *Gravitation*, (W. H. Freeman, San Francisco, 1973); B. S. DeWitt, Phys. Rev. **162**, 1239 (1967), **162**, B1195 (1967); R. P. Feynman, Acta Physica Polonica **24**, 697 (1963); S. Deser and P. van Nieuwenhuizen, Phys. Rev. Letters **32**, 245 (1974); S. Deser, H.-S. Tsao and P. van Nieuwenhuizen, "One Loop Divergences of the Einstein-Yang-Mills System", Brandeis Univ. preprint (1974); S. Weinberg, Phys. Rev. **138**, B988 (1965); L. Smolin, *Three Roads to Quantum Gravity*, (Basic Books, New York, 2001); L. Smolin, "How Far are We From the Quantum Theory of Gravity", (Univ. Waterloo preprint (2003) and references therein; T. Thiemann, "Lectures on Loop Quantum Gravity", Preprint AEI-2002-087, Albert Einstein Insitute, Golm, Germany (2002) and references therein; A. pais and G. E. Uhlenbeck, Phys. Rev. **79**, 145 (1950); G. E. Uhlenbeck, "Lecture Notes on General Relativity", The Rockefeller University (1967), unpublished; S. Blaha, "Generalization of Weyl's Unified Theory to Encompass a Non-Abelian Internal Symmetry Group" SLAC-PUB-1799, Aug 1976; S. Blaha, "Quantum Gravity and Quark Confinement" Lett. Nuovo Cim. **18**, 60 (1977); R. Utiyama, Phys. Rev. **101**, 1597 (1956); T. W. B. Kibble, J. Math. Phys. **2**, 212 (1961); R. Arnowitt, S. Deser, and C. W. Misner, Phys. Rev. **117**, 1595 (1960); and references therein.

We define the proper time differential $d\tau$ as

$$d\tau^2 = g_{\mu\nu}(X(y))dX^\mu dX^\nu \tag{9.1}$$

where, as usual,

$$X^\mu(y) = y^\mu + i\, Y^\mu(y)/M_c^{\,2} \tag{3.12}$$

Thus eq. 9.1 could be written:

$$d\tau^2 = g_{\mu\nu}(X(y))(\eta^\mu{}_\alpha + iM_c^{-2}\partial Y^\mu/\partial y^\alpha)(\eta^\nu{}_\beta + iM_c^{-2}\partial Y^\nu/\partial y^\beta)dy^\alpha dy^\beta \tag{9.2}$$

The inverse of $g_{\mu\nu}$, denoted $g^{\nu\lambda}$, satisfies

$$g_{\mu\nu}(X(y))g^{\nu\lambda}(X(y)) = \delta_\mu{}^\lambda \tag{9.3}$$

Since the algebraic manipulation of the tensor indices is the same as that of the conventional theory of gravitation the two-tier affine connection is:

$${}_X\Gamma^\sigma{}_{\lambda\mu} = \tfrac{1}{2}\, g^{\nu\sigma}\{\partial g_{\mu\nu}/\partial X^\lambda + \partial g_{\lambda\nu}/\partial X^\mu - \partial g_{\lambda\mu}/\partial X^\nu\} \tag{9.4}$$

The two-tier Riemann-Christoffel curvature tensor is:

$${}_XR^\lambda{}_{\mu\nu\kappa} \equiv \partial_X\Gamma^\lambda{}_{\mu\nu}/\partial X^\kappa - \partial_X\Gamma^\lambda{}_{\mu\kappa}/\partial X^\nu + {}_X\Gamma^\alpha{}_{\mu\nu}\,{}_X\Gamma^\lambda{}_{\kappa\alpha} - {}_X\Gamma^\alpha{}_{\mu\kappa}\,{}_X\Gamma^\lambda{}_{\nu\alpha} \tag{9.5}$$

and the two-tier Ricci tensor is

$${}_XR_{\mu\nu} = {}_XR^\alpha{}_{\mu\alpha\nu} \tag{9.6}$$

The two-tier curvature scalar is

$${}_XR = g^{\mu\nu}\,{}_XR_{\mu\nu} \tag{9.7}$$

We also define

$${}_XR_{\lambda\mu\nu\kappa} = g_{\lambda\alpha}\,{}_XR^\alpha{}_{\mu\nu\kappa} \tag{9.8}$$

with the result

$${}_XR_{\lambda\mu\nu\kappa} = \tfrac{1}{2}[\partial^2 g_{\lambda\nu}/\partial X^\kappa\partial X^\mu - \partial^2 g_{\mu\nu}/\partial X^\kappa\partial X^\lambda - \partial^2 g_{\lambda\kappa}/\partial X^\nu\partial X^\mu + \partial^2 g_{\mu\kappa}/\partial X^\nu\partial X^\lambda] +$$

$$+ g_{\alpha\beta}[{}_X\Gamma^{\alpha}{}_{\nu\lambda}\, {}_X\Gamma^{\beta}{}_{\mu\kappa} - {}_X\Gamma^{\alpha}{}_{\kappa\lambda}\, {}_X\Gamma^{\beta}{}_{\mu\nu}] \tag{9.9}$$

We denote the fact that all quantities in eqs. 9.4 – 9.8 are only functions of X by placing a left subscript X on each quantity.

The algebraic properties, and the Bianchi identities, satisfied by ${}_X R_{\lambda\mu\nu\kappa}$ in the two-tier theory of gravitation are identical to those of the conventional theory with all derivatives being with respect to X.

The two-tier version of Einstein's field equations is:

$${}_X R_{\mu\nu} - \tfrac{1}{2}\, g_{\mu\nu}\, {}_X R = -8\pi G\, T_{\mu\nu} \tag{9.10}$$

where G is Newton's gravitational constant ($6.674 \times 10^{-11}\ m^3 kg^{-1} s^{-2}$) and $T_{\mu\nu}$ is the energy-momentum tensor – also strictly a function of X. It is convenient to define the coupling constant

$$\kappa = \sqrt{4\pi G} \tag{9.11}$$

## Lagrangian Formulation

We will now formulate a two-tier Quantum Gravity theory following the same ansatz that we have used throughout this book.

### *Unified Standard Model and Quantum Gravity Lagrangian*

We define the Lagrangian, and action, for the unified quantum field theory of gravitation and the Standard Model as

$$L_{Unified} = \int d^4y\, \mathscr{L}_{Unified} \tag{9.12}$$

$$\mathscr{L}_{Unified} = J\sqrt{g(X)}\, \left(\mathscr{L}_F^{Grav}(X^\mu) + \mathscr{L}_F^{SM}(X^\mu)\right) + \mathscr{L}_C \tag{9.13}$$

with

$$\mathscr{L}_F^{Grav}(X^\mu) = (2\kappa^2)^{-1}\, {}_X R \tag{9.14}$$

where $\mathscr{L}_F^{SM}$ is the complete "normal" Quantum Field theory Lagrangian for the Standard Model version under consideration written in a general covariant form, g(X) is the absolute value of the determinant of $g_{\mu\nu}$, and J is the Jacobian of eq. A.21. *All particle fields in $\mathscr{L}_F^{SM}$ are assumed to be functions of the $X^\mu$ coordinate only. The dependence of the particle fields on the "underlying" coordinates $y^\mu$ is assumed to be* *solely through* $X^{\mu}$. The Lagrangian $\mathscr{L}_{\text{Unified}}$ is a separable Lagrangian of the type of eq. A.26 embodying the composition of extrema described in Appendix A.

As in all of cases that we have considered, we have specified the coordinate part of the Lagrangian $\mathscr{L}_{\text{C}}$ as

$$\mathscr{L}_{\text{C}} = -1/4\, F_{Y}^{\mu\nu} F_{Y\mu\nu} \tag{3.15}$$

with

$$F_{Y\mu\nu} = \partial Y_{\mu}/\partial y^{\nu} - \partial Y_{\nu}/\partial y^{\mu} \tag{3.14}$$

and

$$F_{Y}^{\mu\nu} = \eta^{\mu\alpha}\eta^{\nu\beta} F_{Y\alpha\beta} \tag{9.15}$$

*Why Are the Y Field Dynamics Independent of the Gravitational Field?*

It is evident from eqs. 9.12-3 and 9.15 that the Y field is truly free and, in particular, does not depend on the gravitational field as represented by $\sqrt{g}$ and $g_{\mu\nu}$. Our rationale for this formulation is described in detail in chapter 10. For the moment it suffices to make the following remarks. The Y field is a quantum field at each point in space-time including regions with ultra-strong gravitational fields such as the neighborhoods of black holes. If Y were to depend on the gravitational field then the Y field could be appreciable in such regions and might even be a "classical" field. In this case we would have new dimensions, allbeit imaginary, for which no evidence exists.

Furthermore, the Y part of the Lagrangian establishes a functional relation between the imaginary Y coordinates and the real space-time y coordinates. The Principle of Equivalence applies only to real coordinates and has not been shown to apply to imaginary Quantum Dimensions™. Thus there is no reason to require the Y part of the action to be invariant under general coordinate transformations.

Lastly, the non-invariance of the Y part of the action under general coordinate transformations effectively creates an "absolute" coordinate system – actually a class of "absolute coordinate systems" – namely the class of inertial reference frames that are related to each other by special relativistic transformations. This feature does not conflict with our knowledge of the universe. The universe appears to be almost flat. The large-scale distribution of masses is responsible for this flatness. The flatness, or flattened space if it is slightly curved, together with Mach's Principle (inertial forces are absent in the reference frame determined by the distribution of masses in the universe) selects a preferred class of local reference frames – local inertial reference frames. Since space is almost flat, or flat, these local reference frames occupy a large volume (if we exclude regions with intense gravitational fields.) We can define the Y field within this class of local inertial reference frames in each locale and establish a satisfactory quantum field theory. Thus we have a dynamics defined in the variable X, which we require to be covariant under general coordinate transformations, and a local "ground state" that "breaks" general coordinate invariance down to special relativistic invariance. See chapter 10 for a more complete discussion.

*No "Space-time Foam"*

The fact that our unified theory of the known forces of Nature *self-consistently* has a weak gravitational field at high energies (the graviton sector is finite to all orders in perturbation theory) supports the formulation of eq. 9.12-3. Gravity becomes weaker at ultra-short distances. Therefore space-time is not quantum foam at ultra-short distances but rather smooth and flat a là special relativity – consistent with our formulation.

*Quantum Gravity – Scalar Particle Model Lagrangian*

While the application of the two-tier approach to the unified theory is a straightforward extension of the concepts and approaches described in the preceding chapters, it is useful to consider a simplified model that minimizes the tensorial verbiage so that the concepts and features might better stand out. The procedure differs only in detail from the case of gauge fields.

The introduction of spinor fields requires the use of a two-tier vierbein formalism, which is straightforward to develop. A two-tier vierbein field $e^{\mu}{}_{a}$ is a function of X, $e^{\mu}{}_{a}(X)$, with $g_{\mu\nu} = e_{\mu a}(X)e_{\nu}{}^{a}(X)$ where the index a is an index of a flat tangent space defined at each space-time point. The two-tier formulation of a vierbein theory is similar to the other two-tier formulations that we have considered and will not be developed here.

Thus we will consider the Lagrangian model for a scalar particle field interacting with the $g_{\mu\nu}$ gravitational field:

$$L_{GS} = \int d^4y \sqrt{g(X)}\, \mathscr{L}_{GS} + \mathscr{L}_{C} \tag{9.16}$$

$$\mathscr{L}_{GS} = J\, \mathscr{L}_{F}^{Grav}(X^{\mu}) + J\, \mathscr{L}_{F\phi}(X^{\mu}) \tag{9.17}$$

with covariant versions of eqs. 5.21 and 5.24:

$$\mathscr{L}_{F\phi} = ½\, [\, g^{\mu\nu}\partial\phi/\partial X^{\mu}\, \partial\phi/\partial X^{\nu} - m^2\phi^2] + \mathscr{L}_{F\phi int} \tag{9.18}$$

$$\mathscr{L}_{F\phi int} = 1/4!\, \chi_0\, \phi(X(y))^4 + ½\,(m^2 - m_0{}^2)\phi^2 \tag{9.19}$$

*A Justifiable Weak Field Approximation for Quantum Gravity*

Many discussions of quantizing conventional gravity make a weak field approximation for the gravity sector which, in view of divergences in the resulting quantum field theory, are impossible to justify:

$$g_{\mu\nu} = \eta_{\mu\nu} + \kappa h_{\mu\nu} \tag{9.20}$$

where $\eta_{\mu\nu}$ is the flat space-time metric and $h_{\mu\nu}$ is a "small" deviation ($<h_{\mu\nu}> \ll 1$) from the flat space-time metric.

The two-tier formulation of quantum gravity is finite and the effective field becomes increasingly weaker at short distances. Thus the weak field approximation becomes *more accurate* at short distances:

$$g_{\mu\nu}(X(y)) \simeq \eta_{\mu\nu} + \kappa h_{\mu\nu}(X(y)) \tag{9.21a}$$

At short distances space-time can be considered approximately flat (except possibly in the neighborhood of singularities) with quantum fluctuations embodied in $h_{\mu\nu}$. Thus eq. 9.21a is reasonable within the context of Two-Tier Quantum Gravity.

To first order in $h_{\mu\nu}$ the square root of the absolute value of the determinant of the metric tensor is:

$$\sqrt{g(X)} \simeq 1 + \tfrac{1}{2}\, \kappa h^{\sigma}{}_{\sigma}(X(y)) \tag{9.21b}$$

*Quantization of Quantum Gravity – Scalar Particle Model*

We now proceed to quantize gravity based on the linearization of the gravitational field equations in the weak field approximation. Assuming eq. 9.21a and keeping terms to first order in $h_{\mu\nu}$ gives the affine connection:

$${}_X\Gamma^{\sigma}{}_{\mu\nu} = \tfrac{1}{2}\, \kappa\eta^{\sigma\alpha}[\partial h_{\alpha\nu}/\partial X^{\mu} + \partial h_{\alpha\mu}/\partial X^{\nu} - \partial h_{\mu\nu}/\partial X^{\alpha}] + \mathcal{O}(h^2) \tag{9.22}$$

and the Ricci tensor:

$${}_X R_{\mu\nu} = \partial_X\Gamma^{\lambda}{}_{\lambda\mu}/\partial X^{\nu} - \partial_X\Gamma^{\lambda}{}_{\mu\nu}/\partial X^{\lambda} + \mathcal{O}(h^2) \tag{9.23}$$

Thus the linearized gravitation lagarangian terms are

$$L^{Grav} = \int d^4y\, \sqrt{g(X)}\, J\mathscr{L}_F^{Grav}(X^{\mu}) \rightarrow L^{Grav}{}_{linear} = \int d^4y\, J\mathscr{L}^{Grav}{}_{linear}(X^{\mu}) \tag{9.24}$$

The scalar particle Lagrangian terms become

$$L^{\phi} = \int d^4y\sqrt{g(X)}J\mathscr{L}_{F\phi} \rightarrow \int d^4yJ\{[\tfrac{1}{2}(\eta^{\mu\nu}\partial_{\mu}\phi\partial_{\nu}\phi - m^2\phi^2) + \mathscr{L}_{F\phi int}] +$$

$$+ \tfrac{1}{2}\kappa h^{\mu\nu}\partial_\mu\phi\partial_\nu\phi + \tfrac{1}{4}\,\kappa h(\eta^{\mu\nu}\partial_\mu\phi\partial_\nu\phi - m^2\phi^2) +$$

$$+ \tfrac{1}{2}\,\kappa h\mathscr{L}_{F\phi\text{int}}\} \tag{9.25}$$

with the notation $h = h^\sigma{}_\sigma$ and using

$$\partial_\mu \equiv \partial/\partial X^\mu \tag{9.26}$$

$\eta^{\mu\nu}$ and $\eta_{\mu\nu}$ are used to raise and lower indices in keeping with the linearized, weak field approximation.

The Y terms in the Lagrangian are (as previously):

$$L^Y = \int d^4y\mathscr{L}_C = -\tfrac{1}{4}\int d^4y\eta^{\mu\nu}\eta^{\alpha\beta}F_{Y\mu\alpha}F_{Y\nu\beta} \tag{9.27}$$

We will lump the higher order terms (in h) in the gravity part of the Lagrangian, and the scalar particle part of the Lagrangian, into

$$L_{Higher} = \int d^4y\, J\mathscr{L}_{Higher}(h, \phi) \tag{9.28}$$

Thus the complete lagragian for a scalar particle interacting with gravitons is

$$L_{GS} = L^{Grav}{}_{linear} + L^{\phi}{}_{linear} + L^Y + L_{Higher} \tag{9.29}$$

The linearized gravitational Lagrangian term $L^{Grav}{}_{linear}$ generates the field equations:

$$\Box_X\, h_{\mu\nu} + \partial_\nu\partial_\mu h - \partial_\alpha\partial_\nu h^\alpha{}_\mu - \partial_\alpha\partial_\mu h^\alpha{}_\nu = \kappa S_{\mu\nu} \tag{9.30}$$

where

$$\partial_\mu S^\mu{}_\nu = \tfrac{1}{2}\,\partial_\nu S^\sigma{}_\sigma \tag{9.31}$$

to 0th order in h and where

$$\Box_X = (\partial/\partial X^\nu)(\partial/\partial X_\nu) \tag{9.32}$$

The most general coordinate transformation that maintains the weakness of the gravitational field has the form:

$$y^a \rightarrow \; y'^a = y^a + \chi^a(X(y)) \tag{9.33}$$

This transformation induces a gauge transformation in $h_{\mu\nu}$ to:

$$h'_{\mu\nu} = h_{\mu\nu} - \partial_\mu \chi_\nu - \partial_\nu \chi_\mu \tag{9.34}$$

It is easy to verify that eq. 9.30 is satisfied by $h'_{\mu\nu}$ if it is satisfied by $h_{\mu\nu}$.

Let us assume that we perform a gauge transformation making $h_{\mu\nu}$ traceless:

$$h^\sigma{}_\sigma = 0 \tag{9.35}$$

and choose the gauge

$$\partial^\mu h_{\mu\nu} = 0 \tag{9.36}$$

then eq. 9.30 becomes the wave equation:

$$\Box_X \, h_{\mu\nu} = \kappa S_{\mu\nu} \tag{9.37}$$

Another gauge transformation of the free field $h_{\mu\nu}$ (if $S_{\mu\nu} = 0$) makes

$$h_{\mu 0} = h_{0\mu} = 0 \tag{9.38}$$

while retaining

$$h_{\mu\nu} = h_{\nu\mu} \tag{9.39}$$

The general solution[41] for the free field $h_{\mu\nu}$ (with $S_{\mu\nu} = 0$ in eq. 9.37) can be expressed as a fourier expansion:

$$h_{\mu\nu}(X(y)) = \int d^3k \; N_0(k) \sum_{\lambda=1}^{2} \varepsilon_{\mu\nu}(k, \lambda)[a(k,\lambda)\, e^{-ik\cdot X} + a^\dagger(k,\lambda)\, e^{ik\cdot X}] \tag{9.40}$$

[41] S. Weinberg, Phys. Rev. **135**, B1049 (1964); Phys. Rev. **138**, B988 (1965)

where $\lambda = 1{,}2$ labels the $\pm 2$ helicity states, and where $N_0(k)$ is specified by eq. 3.25. The equal time ($y'^0 = y^0$) commutation relations are:

$$[h_{\mu\nu}(X(y)), h_{\alpha\beta}(X(y'))] = [\pi_{\mu\nu}(X(y)), \pi_{\alpha\beta}(X(y'))] = 0 \tag{9.41}$$

$$[h_{\alpha\beta}(X(y')), \pi_{\mu\nu}(X(y))] = i\, \mathscr{D}_{\alpha\beta\mu\nu}(\mathbf{X}(y) - \mathbf{X}(y')) \tag{9.42}$$

for $\mu$, $\nu = 1, 2, 3$ and where

$$\pi_{\mu\nu}(X(y)) = \partial h_{\mu\nu}(X(y))/\partial y^0 \tag{9.43}$$

in the Y Coulomb gauge where $X^0 = y^0$. $\mathscr{D}_{\alpha\beta\mu\nu}$ is specified by:

$$\mathscr{D}_{\alpha\beta\mu\nu}(X(y) - X(y')) = \int d^3k\, e^{i\,\mathbf{k}\cdot(\mathbf{X}(y) - \mathbf{X}(y'))}\, \Pi_{\alpha\beta\mu\nu}(\mathbf{k})/(2\pi)^3 \tag{9.44}$$

$$\Pi_{\alpha\beta\mu\nu}(\mathbf{k}) = \tfrac{1}{2}\,[(\delta_{\alpha\mu} - k_\alpha k_\mu/\mathbf{k}^2)(\delta_{\beta\nu} - k_\beta k_\nu/\mathbf{k}^2) + (\delta_{\alpha\nu} - k_\alpha k_\nu/\mathbf{k}^2)(\delta_{\beta\mu} - k_\beta k_\mu/\mathbf{k}^2) -$$

$$- (\delta_{\alpha\beta} - k_\alpha k_\beta/\mathbf{k}^2)(\delta_{\mu\nu} - k_\mu k_\nu/\mathbf{k}^2)] \tag{9.45}$$

where $\alpha$, $\beta$, $\mu$, $\nu = 1, 2, 3$.

The "transverse" graviton propagator can be represented as a time-ordered product of field operators:

$$i\Delta_{F2}{}^{TT}(y_1 - y_2)_{\lambda\tau\rho\sigma} = \langle 0|\,T(h_{\lambda\tau}(X(y_1)), h_{\rho\sigma}(X(y_2)))\,|0\rangle \tag{9.46}$$

$$= -i \int \frac{d^4k\, e^{-ik\cdot(y_1 - y_2)}\, b_{\lambda\tau\rho\sigma}(k) R(\mathbf{k}, y_1 - y_2)}{(2\pi)^4\,(k^2 + i\varepsilon)}$$

where $R(\mathbf{k}, y_1 - y_2)$ is the gaussian factor appearing in propagators throughout two-tier theories, $\mathbf{k}$ is a spatial 3-vector, and where $b_{\mu\nu\rho\sigma}(k)$ is a function of k only:

$$b_{\alpha\beta\mu\nu}(k) = \tfrac{1}{2}[(\eta_{\alpha\mu} - k_\alpha k_\mu/\mathbf{k}^2)(\eta_{\beta\nu} - k_\beta k_\nu/\mathbf{k}^2) + (\eta_{\alpha\nu} - k_\alpha k_\nu/\mathbf{k}^2)(\eta_{\beta\mu} - k_\beta k_\mu/\mathbf{k}^2) -$$

$$- (\eta_{\alpha\beta} - k_\alpha k_\beta/\mathbf{k}^2)(\eta_{\mu\nu} - k_\mu k_\nu/\mathbf{k}^2)] \qquad (9.47)$$

where $\alpha, \beta, \mu, \nu = 0, 1, 2, 3$.

The quantum gravitational interaction also has an "instantaneous" part (similar to the instantaneous Coulomb interaction of QED) in addition to the transverse interaction embodied in eq. 9.46. This "instantaneous" interaction contains the Newtonian potential (described later) as its large distance limit. The sum of the instantaneous interaction and the transverse interaction gives the total gravitational interaction.

The above graviton propagator has the form given in eq. 6.102. The caculation of the leading behavior is the same as that of the two-tier scalar boson propagator except for the presence of factors such as $\eta_{\rho\sigma}$. The leading momentum dependence of the graviton propagator in momentum space is

$$i\Delta_{F2}{}^{TT}(p)_{\lambda\tau\rho\sigma} \backsim p^{-6} \qquad (9.48)$$

The graviton vertices in two-tier Quantum Gravity will be described within the framework of the path integral formulation.

*Quantum Gravity–Scalar Particle Model Path Integral*

A path integral formalism can be developed for two-tier Quantum Gravity interacting with matter fields. In this section we will consider the case of a matter field consisting of massive scalar bosons with a quartic interaction. The path integral formalism that we develop is similar to that of Yang-Mills theories in the previous chapter.

The two-tier path integral for a Quantum Gravity–Scalar Particle Theory can be written as:

$$Z(J, J^{\mu\nu}) = N \int D\phi Dh DY \Delta_{FPG}(h)\delta(F(h)) \exp\{i\int d^4y[\mathscr{J}(\mathscr{L}^{Grav}{}_{linear}(X^\mu) +$$

$$+ \mathscr{L}^{\phi}{}_{linear}(X^\mu) + \mathscr{L}_{Higher}(h, \phi)) + \mathscr{L}_C(X, y) +$$

$$+ j_\mu(y)Y^\mu(y) + J(y)\phi(X) + J^{\mu\nu}(y)h_{\mu\nu}(X)]\}\Big|_{j_\mu = 0} \qquad (9.49)$$

where $\delta(F(h))$ specifies the gauge as a functional delta function, and $\Delta_{FPG}(h)$ is the corresponding Fadeev-Popov determinant. $\mathscr{J}$ is the Jacobian for the transformation from y coordinates to X coordinates. The Fadeev-Popov determinant $\Delta_{FPG}(h)$ can be calculated in the standard way. First we note

$$\delta(F(h^{\chi})) = \delta(\chi - \chi_0) \ \left| \det \delta F(h_{\mu\nu}{}^{\chi}(X))/\delta\chi(X) \right|^{-1} \Big|_{F(h)=0} \tag{9.50}$$

where

$$h_{\mu\nu}{}^{\chi} = h_{\mu\nu} - \partial_{\mu}\chi_{\nu} - \partial_{\nu}\chi_{\mu} \tag{9.34}$$

Then

$$\Delta_{FPG}(h) = \ \left| \det \delta F(h^{\chi}(X))/\delta\chi(X) \right| \Big|_{F(h)=0} \tag{9.51}$$

We will choose the gauge of eq. 9.36 to evaluate the Fadeev-Popov determinant. Under an infinitesimal gauge transformation of the form:

$$h_{\mu\nu}{}^{\chi}(X) = h_{\mu\nu}(X) - \partial_{\mu}\chi_{\nu} - \partial_{\nu}\chi_{\mu} \tag{9.52}$$

which preserves the weak field nature of $h_{\mu\nu}$, we find

$$F_{\nu}(h^{\chi}) = \partial^{\mu}\,(h_{\mu\nu}(X) - \partial_{\mu}\chi_{\nu} - \partial_{\nu}\chi_{\mu})$$

$$= - \Box_X\, \chi_{\nu}(X) - \partial_{\nu}\partial^{\mu}\,\chi_{\mu} \tag{9.53}$$

Thus

$$\delta F_{\mu}(h^{\chi}(X))/\delta\chi^{\nu}(X) = - \eta_{\mu\nu}\Box_X - \partial_{\mu}\partial_{\nu} \tag{9.54}$$

and

$$\Delta_{FP}(A) = \ \left| \det\,( - \eta_{\mu\nu}\Box_X - \partial_{\mu}\partial_{\nu}) \right| \Big|_{F(h)=0} \tag{9.55}$$

We note the two-tier Fadeev-Popov determinant is solely a function of the X coordinates. The determinant only introduces an overall multiplicative constant that can be absorbed into the normalization constant N. This fact becomes evident if we follow the standard procedure and rewrite the determinant as a path integral over anti-commuting c-number fields with a ghost Lagrangian. Then we see that the ghost does not interact with the other fields and thus only generates an overall multiplicative constant that can be absorbed in N:

$$\Delta_{FPG}(h) = \int Dc^{*}Dc \exp[\, i\int d^4X\, \mathscr{L}^{ghost}(X^{\mu})] \tag{9.56}$$

where

$$\mathscr{L}^{ghost}(X^{\mu}) = c^{\mu *}(X)[\eta_{\mu\nu}\Box_X + \partial_{\mu}\partial_{\nu}]c^{\nu}(X) \tag{9.57}$$

We now go through the same analysis as we did in the $\phi^4$ theory path integral example and the Yang-Mills path integral example (with some superficial differences). First we integrate the linear part of the Y field Lagrangian as we did previously. Then we integrate the linear part of the $\phi$ field Lagrangian as done previously. Lastly we integrate the linear part of the gravitation Lagrangian to obtain the path integral for the perturbative expansion with the result:

$$Z(J, J^{\mu\nu}) = N \left\{\exp\left[i\int d^4y \mathscr{L}_{Higher}(\partial/\partial y^{\nu}, -i\delta/\delta J^{\mu\nu}(y), -i\delta/\delta J(y))\right]\cdot\right.$$

$$\cdot \exp[-\tfrac{1}{2}\, i\int d^4y_1 d^4y_2 J^{\mu\nu}(y_1)\Delta_{F2}{}^{TT}(y_1 - y_2, z)_{\mu\nu\rho\sigma}J^{\rho\sigma}(y_2)]\cdot$$

$$\left.\cdot \exp[-\tfrac{1}{2}\, i\int d^4y_1 d^4y_2 J(y_1)\Delta_F{}^{TT}(y_1 - y_2, z)J(y_2)]\right\} \Big|_{z=y_1-y_2} \tag{9.58}$$

There are two issues that arise in the development of eq. 9.58:

1.) The integral over y in $\int d^4y\mathscr{L}_{Higher}$ which began as the integral $\int d^4y \mathscr{J} \mathscr{L}_{Higher} = \int d^4X\mathscr{L}_{Higher}$ in eq. 9.49; and

2.) The handling of derivatives with respect to X in $\mathscr{L}_{Higher}$.

These are resolved by the following respective observations:

1.) See the discussions following eqs. 6.34 and 8.29 that apply here as well without change.

2.) See the discussion following eq. 8.32, which applies here with only superficial changes. In particular we note that the derivative with respect to X of the graviton propagator (eq. 9.46-7) is specified by the following:

$$\partial i\Delta_{F2}{}^{TT}(y_1 - y_2)_{\lambda\tau\rho\sigma}/\partial X^\mu(y_1) = \partial[i\Delta_{F2}{}^{TT}(y_1 - y_2, z)_{\lambda\tau\rho\sigma}]/\partial y_1{}^\mu \Big|_{z = y_1 - y_2} \tag{9.59}$$

where

$$i\Delta_{F2}{}^{TT}(y_1 - y_2, z)_{\lambda\tau\rho\sigma} = -i\int \frac{d^4k\, e^{-ik\cdot(y_1 - y_2)}\, b_{\lambda\tau\rho\sigma}(k)R(\mathbf{k}, z)}{(2\pi)^4\,(k^2 + i\varepsilon)} \tag{9.60}$$

Thus

$$\frac{\partial\, i\Delta_{F2}{}^{TT}(y_1 - y_2)_{\lambda\tau\rho\sigma}}{\partial X^\mu(y_1)} = -i\int \frac{d^4k\, e^{-ik\cdot(y_1 - y_2)}\,(-ik_\mu)b_{\lambda\tau\rho\sigma}(k)R(\mathbf{k}, y_1 - y_2)}{(2\pi)^4\,(k^2 + i\varepsilon)} \tag{9.61}$$

Therefore derivatives with respect to X in the interaction Lagrangian terms can be replaced by derivatives with respect to y if the graviton propagator is generalized to eq. 9.60. After taking all derivatives with respect to y, we set z equal to the respective $y_1 - y_2$ (actually the difference of the appropriate variables) in each propagator with results similar to eq. 9.61.

$$\begin{aligned} Z(J, J^{\mu\nu}) = N\, \{&\exp[i\textstyle\int d^4y\,\mathscr{L}_{Higher}(\partial/\partial y^\nu, -i\delta/\delta J^{\mu\nu}(y), -i\delta/\delta J(y))]\cdot \\ &\cdot \exp[-\tfrac{1}{2}\, i\textstyle\int d^4y_1 d^4y_2 J^{\mu\nu}(y_1)\Delta_{F2}{}^{TT}(y_1 - y_2, z)_{\mu\nu\rho\sigma}J^{\rho\sigma}(y_2)]\cdot \\ &\cdot \exp[-\tfrac{1}{2}\, i\textstyle\int d^4y_1 d^4y_2 J(y_1)\Delta_F{}^{TT}(y_1 - y_2, z)J(y_2)]\}\Big|_{z=y_1 - y_2} \end{aligned} \tag{9.58a}$$

To be precise eq. 9.58a is interpreted as executing the following steps:

1. For a given process take appropriate functional derivatives of Z(J) with respect to J and $J^{\mu\nu}$.

2. Then expand the exponential factors in a perturbation series applying any derivatives with respect to y in $\mathscr{L}_{\text{Higher}}$. Do not perform any of the $\int d^4y_1 d^4y_2$ integrals.

3. Then set $z = y_1 - y_2$ in each $\Delta_{Fk}{}^{TT}(y_1 - y_2, z)$ and $\Delta_{F2}{}^{TT}(y_1 - y_2, z)_{\mu\nu\rho\sigma}$ propagator.

4. Lastly perform all $\int d^4y_1 d^4y_2$ integrals.

Thus we achieve a path integral formulation that is very similar to the corresponding expression in conventional field theory – the only difference is in the form of the free field propagators, which each now contain a Gaussian factor. The net consequence is that graviton vertices result in exactly the same polynomials in momenta as the conventional theory.

Thus two-tier gravity generates a perturbative expansion identical to conventional quantum gravity except that each graviton propagator has a gaussian damping factor $R(\mathbf{k}, y_1 - y_2)$. At low energies the tree diagrams of conventional gravity theory emerge to good approximation in two-tier gravity. All diagrams with loops converge. Thus two-tier gravity is finite.

*Finiteness of Quantum Gravity–Scalar Particle Model*

Two-tier Quantum Gravity perturbation theory is finite. Calculations are highly convergent at large momentum ($\gtrsim M_c$). At low momentum the two-tier theory is similar to conventional gravity – particularly for tree diagrams and other convergent diagrams in conventional quantum gravity.

For pure *conventional* Quantum Gravity DeWitt[42] finds the superficial degree of divergence of a diagram to be:

$$D = -2L_i + 2\sum_n V_n + 4K \qquad (9.62)$$

where $L_i$ is the number of internal lines, $V_n$ is the number of n-pronged vertices, and K is the number of independent momentum integrations. DeWitt further points out

$$K = L_i - \sum_n V_n + 1 \qquad (9.63)$$

Thus the superficial degree of divergence of a <u>*conventional*</u> Quantum Gravity diagram is:

$$D = 2(K + 1) \qquad (9.64)$$

---

[42] B. S. DeWitt, Phys. Rev. **162**, 1239 (1967).

for $K \geq 1$, displaying an ever increasing degree of divergence as the order of the diagram increases.

In the case of *Two-Tier Quantum Gravity* the superficial degree of divergence of a diagram is:

$$D_{TT} = -6L_i + 2\sum_n V_n + 4K \tag{9.65}$$

(from eq. 9.48) with the result (taking account of eq. 9.63):

$$D_{TT} = -2L_i - 2\sum_n V_n + 2 \tag{9.66}$$

Since any diagram with a loop has $L_i \geq 1$ and $\sum_n V_n \geq 1$ we see that $D \leq -2$. Thus *all* diagrams are convergent and *the two-tier formulation of Quantum Gravity theory is finite. The addition of arbitrary species of other two-tier fields – matter and gauge fields – does not introduce divergences in the combined two-tier theory.*

*Unitarity of Quantum Gravity–Scalar Particle Model*

The two-tier Quantum Gravity – Scalar Particle Model *superficially* appears to have a unitarity problem due to the non-hermitean nature of its hamiltonian. The lack of hermiticity is due entirely to the appearance of $iY^\mu$ in the $X^\mu$ field coordinates.

Thus interaction Lagrangian is not hermitean:

$$L_{Higher} = \int d^3y' \mathscr{L}_{Higher}(\, y' + iY(y')/M_c^2) \tag{9.67}$$

and

$$L_{Higher}^{\dagger} = \int d^3y' \mathscr{L}_{Higher}(\, y' - iY(y')/M_c^2) \neq L_{Higher} \tag{9.68}$$

The relation between $L_{Higher}$ and its hermitean conjugate is

$$L_{Higher} = V\, L_{Higher}^{\dagger}\, V \tag{9.69}$$

where $V^2 = I$ is the metric operator defined in eqs. 5.16 – 5.18. By eq. 6.37 we see as a result that the two-tier S matrix is not unitary – it is pseudo-unitary:

$$S^{-1} = V\, S^{\dagger}\, V \tag{9.70}$$

Therefore

$$S^{\dagger} V S = V \tag{9.71}$$

The S matrix satisfies the unitarity condition between physical asymptotic states – states consisting of only scalar $\phi$ particles and gravitons. The proof is identical in form to eqs. 6.46 – 6.48. The S matrix of the unified theory of the Standard Model and Quantum Gravity can be similarly shown to satisfy the unitarity condition.

## The Mass Scale $M_c$

The mass scale of two-tier theories is set by $M_c$. This mass scale cannot be ascertained with any degree of certainty at current, experimentally accessible, accelerator energies. Cosmic ray data also does not seem to give any clues as to the value of $M_c$. It appears that $M_c$ is probably above $10^3$ GeV/c$^2$ and may be of the order of (or equal to) the Planck mass:

$$M_{planck} = \sqrt{\hbar c/G} = 1.22\times10^{19}\ GeV/c^2 \tag{9.75}$$

If $M_c$ is of the 1,000 GeV/c$^2$ or larger the differences between its predictions at current accelerator energies and the predictions of conventional renormalized perturbation theory will be negligible. Actually a much lower value of $M_c$ would still be consistent with the current stringent QED theoretical predictions as well as other predictions of conventional renormalized perturbation theory.

## Planck Scale Physics

A finite theory of Quantum Gravity can provide information on the issues that have been of concern for many years – including the short distance behavior of the gravitational metric and ultra-small black holes.

### *Quantum Foam*

Some theorists have conjectured that the classical view of smooth, almost flat space-time does not hold in the quantum regime at energies of the order of the Planck mass. Suggestions that space-time dissolves into quantum foam have appeared.

The finite two-tier formulation of Quantum Gravity is well-behaved at short distances and suggests that the quantum behavior of gravity and space-time in the short distance limit does not have limitless quantum fluctuations that result in a foam-like space-time picture.

## Measurement of the Quantum Gravity Field

A number of conceptual problems have been raised about the effects of quantized General Relativity. Two-tier Quantum Gravity seems to resolve these issues.

*Measurement of Time Intervals*

Wigner[43] has studied the measurement of time intervals in General Relativity and sees a problem in the measurement of extremely short intervals. According to Wigner: the measurement of a time inteval in a region of space requires the measurement of the length of time required for an event to happen. The measurement requires an accurate clock. But the accuracy of the clock is limited by the energy-time uncertainty relation:

$$\Delta E \Delta t \geq \hbar \tag{9.76}$$

Thus the uncertainty in the clock's time measurement is related to the uncertainty in the clock's energy which is, in turn, related to the uncertainty in the clock's mass:

$$\Delta E = (\Delta m)c^2 \tag{9.77}$$

To obtain "infinite" accuracy the uncertainty (fluctuations) in the clock's mass must be infinite and thus the clock's mass must be infinite. Infinite fluctuations in the clock's mass will produce corresponding infinite fluctuations in the gravitational field.

$$\Delta h \propto \Delta E \qquad \text{(in conventional General Relativity)} \tag{9.78}$$

As a result the notion of space-time and time intervals (which depend on the geometry through General Relativity) become uncertain. Thus, according to Wigner and others, the concept of time intervals and space-time points becomes questionable.

The two-tier version of Quantum Gravity offers a potential way out of this dilemma. The gravitational force becomes stronger as one goes to shorter distances (higher energies) down to a distance (up to an energy) whose scale is set by $M_c$. At shorter distances (higher energies) the gravitational force becomes weaker and declines to zero at zero distance. Thus at very high energy the gravitational field fluctuations ($\Delta h$) are at worst inversely proportional to the energy (and probably decline by a higher power of inverse energy.) (The same considerations would apply if one chooses to consider fluctuations in the Riemann-Christoffel symbols.)

$$\Delta h < c_1/E < c_1/(\Delta E) \qquad \text{(in two-tier Quantum Gravity)} \tag{9.79}$$

43 E. P. Wigner, Rev. Mod. Phys. **29**, 255 (1957); J. Math. Phys. **2**, 207 (1961).

Thus Wigner's conclusion does not hold in the two-tier version of Quantum Gravity as gravitational fluctuations actually become smaller at energies above a critical energy whose scale is set by $M_c$.

In fact, combining eqs. 9.79 and 9.76 we see

$$c_1 \Delta t / \Delta h \geq \hbar \tag{9.80}$$

at sufficiently high energy. Therefore the time uncertainty $\Delta t$, and the gravitational field fluctuations $\Delta h$, can both decrease while maintaining the energy-time uncertainty relation. *Thus the notion of a space-time point "is saved" in two-tier quantum gravity.*

*Vacuum Fluctuations in the Gravitation Fields*

While the expectation value of the free graviton field $h_{\mu\nu\text{conv}}(X)$ is zero in a conventional quantum field theoric approach:

$$<0|\, h_{\mu\nu\text{conv}}(X)\,|0> = 0 \tag{9.81}$$

the vacuum fluctuations of the *conventional* quantum graviton field is quadratically divergent since

$$<0|\, h_{\mu\nu\text{conv}}(X) h_{\alpha\beta\text{conv}}(X)\,|0> = \int d^3p\; b'_{\mu\nu\alpha\beta}(p)/[(2\pi)^3\, 2\omega_p] = \infty \tag{9.82}$$

where $b'_{\mu\nu\alpha\beta}(p)$ is a rational function of the momentum p.

In "two-tier" quantum field theory we find

$$<0|\, h_{\mu\nu}(X) h_{\alpha\beta}(X)\,|0> = \int d^3p\; b'_{\mu\nu\alpha\beta}(p)\; e^{-p^i p^j \Delta_{Tij}(0)}/[(2\pi)^3 2\omega_p] = 0 \tag{9.83}$$

since the exponential factor in the integrand is $-\infty$. The exponent contains

$$\Delta_{Tij}(z) = \int d^3k\; e^{-ik\cdot z}(\delta_{ij} - k_i k_j/\mathbf{k}^2)/[(2\pi)^3 2\omega_k] \tag{4.8}$$

Thus the vacuum fluctuations of $h_{\mu\nu}$ are zero in "two-tier" quantum field theory.

## The Two-Tier Gravitational Potential vs. Newton's Gravitational Potential

The familiar gravitational potential of Newton is:

$$V_{\text{Newton}} = -\,G/\,|\mathbf{r}| \tag{9.84}$$

The Two-Tier gravitational potential is:

$$V_{\text{Two-Tier}} = -\,G\Phi(M_c^{\,2}\pi\,|\mathbf{r}|^2)/\,|\mathbf{r}| \tag{9.85}$$

where $\Phi(y)$ is the error function.[44] It can be calculated in Two-Tier Quantum Gravity from Two-Tier Quantum Gravity propagator terms similar to corresponding terms in the Two-Tier photon propagator that led to the Two-Tier Coulomb potential (eqs. 7.48 – 7.51). At small distances ($\pi r^2 \ll M_c^{-2}$)

$$V_{\text{Two-Tier}} \rightarrow -\,G2\surd\pi\;M_c^{\,2}\,|\mathbf{r}| \tag{9.86}$$

a linear potential, and at large distances ($\pi r^2 \gg M_c^{-2}$)

$$V_{\text{Two-Tier}} \rightarrow V_{\text{Newton}} = -\,G/\,|\mathbf{r}| \tag{9.87}$$

the Newtonian potential.

The Two-Tier gravitational potential has a minimum at

$$M_c^{\,2}\pi r_{\text{MIN}}^{\;2} = 1 \tag{9.88}$$

At the minimum $V_{\text{Two-Tier}}$ has the value:

$$V_{\text{Two-TierMIN}} = -\,.8427G\surd\pi\;M_c \tag{9.89}$$

Figs.9.1 – 9.2 display plots of $V_{\text{Two-Tier}}$ for $M_c$ = 1 TeV/$c^2$, and $M_c$ = 1.22 $10^{19}$ GeV/$c^2$ = $G^{-1/2}$ – the Planck mass.

---

44 W. Magnus and F. Oberhettinger, *Formulas and Theorems for the Special Functions of Mathematical Physics* (Chelsea Publishing Co., New York, 1949) page 96.

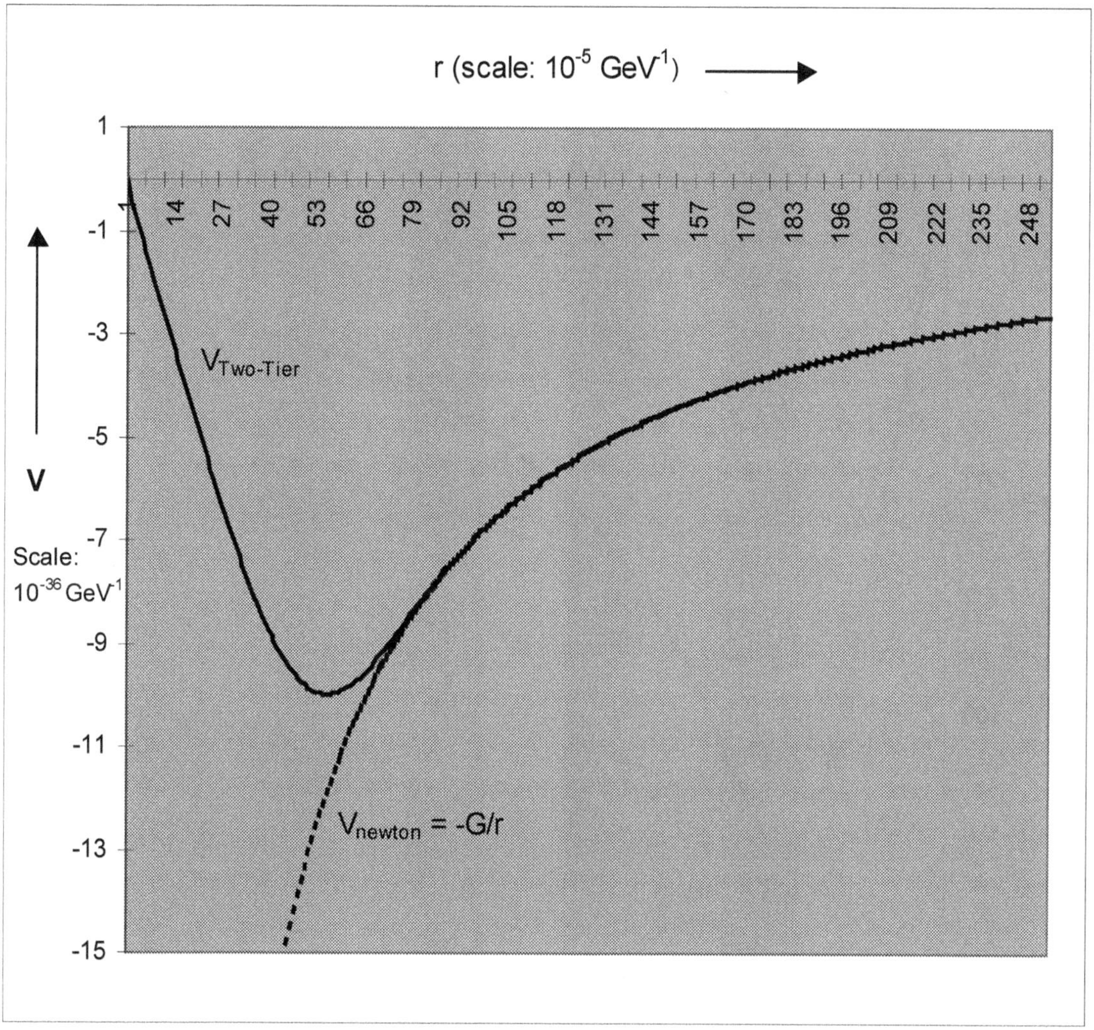


Figure 9.1. Plot of Two-Tier gravitational potential for $M_c$ = 1 TeV/c$^2$ and Newton's gravitational potential. The potentials are measured in units of $10^{-36}$ GeV$^{-1}$. The radial distance is measured in units of $10^{-5}$ GeV$^{-1}$. The plot of the two-tier potential shows the force of gravity is repulsive for small $r < 5.7 \times 10^{-4}$ GeV$^{-1}$.

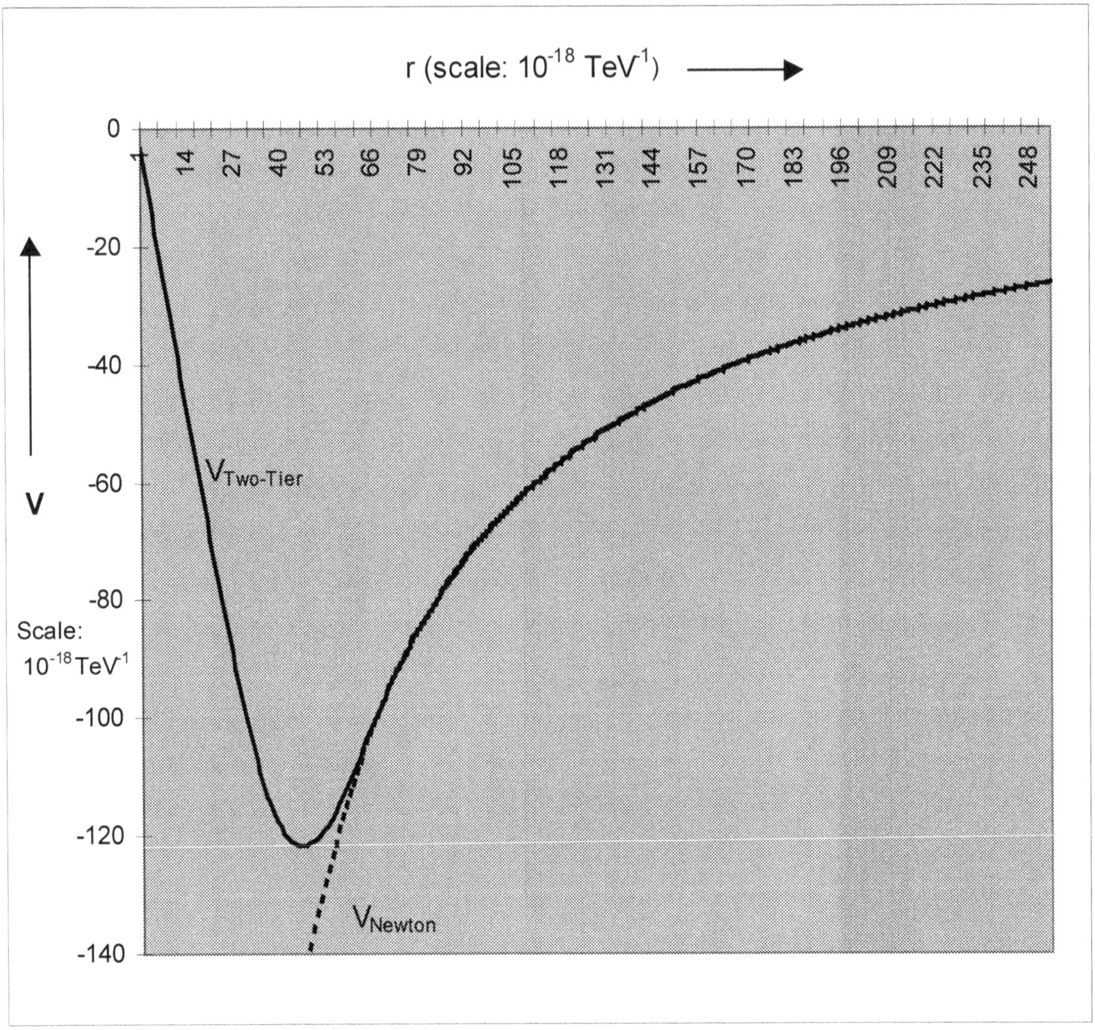


Figure 9.2. Plot of Two-Tier gravitational potential for $M_c = 1.22 \times 10^{19}$ GeV/c$^2$ (the Planck mass) and Newton's gravitational potential. Potentials are measured in units of $10^{-18}$ TeV$^{-1}$. The radial distance is measured in units of $10^{-18}$ TeV$^{-1}$.

## Black Holes

The existence of microscopic black holes has been the subject of much speculation. It appears that arbitrarily small black holes can exist in classical General Relativity. The divergences associated with the short distance behavior of its conventional quantization raise the possibility of additional singular behavior at short distances as well.

On the other hand, in two-tier Quantum Gravity, at short distances, when the distance scale becomes less than $M_c^{-1}$ (and thus the energy scale becomes greater than $M_c$), the two-tier gravitational force grows smaller and become zero in the limit of zero distance or infinite energy. The preceding figures (Figs. 9.1 – 9.2) show the two-tier gravitational potential linearly approaches zero at short distances unlike the Newtonian gravitational potential which approaches $-\infty$ as r approches zero. (The transverse gravitational propagator also approaches zero at short distances.) Thus the short distance behavior of two-tier gravity suggests that black holes of ultra-small size may not exist in two-tier Quanum Gravity.

If we examine the two-tier gravitational potential we note that it is similar to the Newtonian potential until the separation distance approaches the minimum of the potential. Thus we might expect that conventional classical General Relativity would be approximately valid down to distances of the order of the location of the minimum of the two-tier potential. Based on this assumption and on the assumption that $M_c$ equals the Planck mass:

Assumption: $$M_c = M_{Planck} = G^{-1/2} \tag{9.90}$$

we can calculate the mass of a black hole whose radius equals the minimum of the two-tier potential. From eq. 9.88 we obtain

$$r_{MIN} = (G/\pi)^{1/2} = r_{BlackHole} = 2GM_{BlackHoleMIN} \tag{9.91}$$

with the result

$$M_{BlackHoleMIN} = (4\pi G)^{-1/2} = \kappa^{-1} \tag{9.92}$$

by eq. 9.11 and

$$M_{BlackHoleMIN} \cong .282\ M_{Planck} \tag{9.93}$$

or $6.15 \times 10^{-6}$ grams. This lower limit on black hole mass is substantially greater than the collision energy than can be achieved in any current particle accelerator. Thus the production of ultra-small black holes in particle accelerators is unlikely.

Since corrections to conventional quantum gravity are at most of the order of $M_c^{-2}$ it appears that the value of $M_{BlackHoleMIN}$ is consistent with the approximate validity of classical expression for a black hole radius. We note

$$(M_{BlackHoleMIN}/M_c)^2 \cong .0795 \tag{9.94}$$

and so corrections to eq. 9.93 would be very small.

# 10. Curved Space-time Generalization of Two-tier Quantum Gravity

## Inertial Reference Frames & Absolute Space-time

The concept of a flat, absolute space-time can be defined in two ways:

1. There exists a specific reference frame, an *absolute reference frame*, with space-time coordinates that we will denote as $y^{\mu}$ for $\mu = 0, 1, 2, 3$. Any reference frame whose space-time coordinates $y'^{\mu}$ are related to the $y^{\mu}$ coordinates by equations of the form:

$$y'^{j} = R^{j}_{i}y^{i} + v^{j}y^{0} + c^{j} \tag{10.1}$$

$$y'^{0} = y^{0} + c^{0} \tag{10.2}$$

where $R^{j}_{i}$ is a constant, real, orthogonal matrix, and where $v^{j}$ and $c^{\mu}$ are constants for $j = 1, 2, 3$ and $\mu = 0, 1, 2, 3$ is an equivalent inertial reference frame. The set of these reference frames is called the set of *inertial reference frames*. The form of the equations of motion for a set of point-like particles is the same in any inertial reference frame. Thus these reference frames are physically equivalent.

2. There is a class of reference frames called inertial reference frames whose coordinates are related by equations of the form of eqs. 10.1 and 10.2. The form of the equations of motion for a set of point-like particles is the same in any inertial reference frame and for slowly moving particles have the form of Newton's equations without inertial forces. No inertial reference frame has any special significance.

Current cosmological data suggest that space is almost flat, or flat. Thus we can establish either local inertial reference frames (curved space case) or global inertial reference frames (flat space case) as experiment eventually will indicate.

Since the form of the equations of motion is the same in all local inertial frames it would appear that there is no way to physically distinguish between definitions 1 and 2. However, the observed characteristics of cosmic background radiation (CBR), of the redshift – distance relationship, and of the cosmological X-ray background effectively define a preferred local reference frame in each spatial locale. Effectively the CBR plays the role of an *aether selecting a preferred local inertial reference frame in each spatial locale.* The set of all such preferred inertial reference frames for the universe effectively defines an Absolute Reference Frame. IF space is truly flat, then the Absolute Reference Frame consists of one inertial reference frame.

The views of physicists on the question of an absolute reference frame have oscillated over time. Newton chose the first definition and talked of "absolute space" in the famous quote, "Absolute space in its own nature, and with regard to anything external, always remains similar and immovable." Mach challenged Newton's view and enunciated what became known as Mach's Principle postulating that there existed a class of inertial frames (second type of definition) that were defined by the mass distribution, and its movement, of the universe. Einstein established an "intermediate" position between Newton and Mach in his General Theory of Relativity. General Relativity implicitly defines the equivalent of an absolute space and shows that the presence of masses is not required since inertial frames exist in the empty space solutions of the equations of general relativity.

However the dynamical equations of General Relativity are covariant under any general relativistic transformation. *One can view General Relativity as a theory embodying invariance under general relativistic transformations with the invariance "broken" by a class of equivalent local "ground states" called local inertial reference frames that are determined by Mach's Principle, and eqs. 10.1 and 10.2 – an analogue of spontaneous broken symmetry.*

In this chapter we shall examine the Newtonian, Machian, and Einsteinian views in more detail and discuss them in relation to two-tier quantum field theory.

## Newtonian Mechanics Embodies an Absolute Space-time

Newton developed his formulation of mechanics with equations of motion for groups of point-like particles. These equations had the same form in all inertial reference frames. He then asserted that the class of inertial reference frames was selected because they were either at rest or in a state of constant velocity with respect to a particular reference frame that corresponded to *absolute space*. He postulated the existence of a *physical* absolute space partly for theological reasons.

## Mach's Principle Embodies an Absolute Space-time

Ernst Mach disagreed with Newton's concept that the basis for the special properties of inertial reference frames was their relation to the reference frame of an absolute space. Following Leibniz and others he proposed another view that is now called *Mach's Principle*.

An example[45] that illustrates Mach's thinking is:

> Consider a universe consisting of two identical spheres, not necessarily in close proximity, upon each of which an ant is stationary on the sphere's equator. Assume the spheres are rotating with respect to each other with parallel axes of rotation. The ant on each sphere sees the ant on the other sphere rotating with the sphere. Questions: which ant experiences an "upward" centrifugal force? Which ant experiences a Coriolis force (which is proportional to the angular velocity of rotation)?

These questions do not have answers in the universe of two spheres subject only to Newton's laws according to Ernst Mach. Either sphere could be considered to be the non-rotating sphere and a corresponding valid coordinate system defined in which the other sphere would be rotating.

Mach resolved this issue by noting the universe is very large and populated with large masses in all directions at great distance. In his view the distribution and motion of all the masses in the universe defined a preferred reference frame, although it is not usually portrayed in that manner. Mach would then have resolved the two spheres issue by saying the rotation of each sphere must be determined relative to the distribution and motion of the rest of the matter in the *real* universe.

As R. H. Dicke has remarked,[46] "If one were to remove this matter [at great distances], then according to Mach, the inertial force would disappear. If one were to reduce the matter to negligible proportions, there would be striking changes in local inertial effects. To summarize, according to Mach's point of view, we should interpret inertial effects as a consequence of interactions of matter at great distances in the universe with accelerated bodies in the laboratory."

Although many physicists, including Einstein, were strongly influenced by Mach's arguments many physicists were also uneasy about Mach's Principle. As Eddington[47] remarked, on the use of matter at infinity to define inertial frames, "the main feeling seems to be that it is unsatisfactory to have certain conditions prevailing in the world, which can be traced away to infinity and so have, as it were, their source at infinity; and there is a desire to find some explanation of the inertial frame as built up through conditions at a finite distance."

Thus Mach's Principle was viewed with mixed feelings both before, and long after, Einstein's development of his general theory of relativity.

---

45 Mach (1991).

46 R. H. Dicke, lecture entitled *The Many Faces of Mach* (1963).

47 Eddington (1995) p. 157.

## General Relativity Embodies an Absolute Space-time

Einstein, himself, was strongly influenced by Mach's arguments. Mach's Principle was certainly on his mind when he was formulating his equivalence principle and the general theory of relativity. To some extent he felt his theory embodied Mach's Principle. However in the view of almost all observers General Relativity only partly implements Mach's Principle. General Relativity *also* embodies aspects of an absolute space-time.

R. H. Dicke,[48] points out:

> "Einstein's theory is not relativistic in the Machian sense. In his theory, space has physical properties and constitutes a physical structure even in the absence of all matter. ... general relativity does not appear to describe Mach's principle properly. This can be seen by noting that, in the absence of all matter, the metric tensor describes a flat space and this flat space possesses inertial properties. Even Schwarzschild's famous solution is unsatisfactory, from the point of view of Mach. As one moves to infinity, and the mass source (the source of inertial forces according to Mach) disappears in the distance, the space becomes flat and continues to possess inertial properties in contradiction with the expectations of Mach. ...
>
> We have ... the return to the idea that we are dealing with an absolute space-time. From the viewpoint of Synge, general relativity describes the geometry of an absolute space. According to him, certain things are measurable about this space in an absolute way. There exist curvature invariants that characterize this space, and one can, in principle, measure these invariants. Bergmann has pointed out that the mapping of these invariants throughout space is, in a sense, a labeling of the points of this space with invariant labels (independent of coordinate system). These are concepts of an absolute space, and we have here a return to the old notions of an absolute space."

## The Case for Absolute Space-time

With the very strong case for absolute space-time made by Synge, Dicke and Bergmann, three of the great general relativists, we now ask whether a return to absolute space-time is in order in the sense of definition 2 above.

The arguments for an absolute space-time are:

1. It is embodied in the two successful theories of mechanics: Newtonian mechanics and general relativity.

2. It supports a local definition of physics consistent with the spirit of Riemannian geometry, general relativity and quantum field theory.

[48] R. H. Dicke, lecture entitled *The Many Faces of Mach* (1963).

3. **Experimental data is consistent with the existence of absolute rotation. As Eddington[49] notes, "The great stumbling block for a philosophy which denies absolute space is the experimental detection of absolute rotation." Most interestingly, current cosmological experimental data suggests space is very close to flat if not flat. Thus we are living in an absolute reference frame if the universe is considered in the large (with local masses averaged over cosmological distances.)**

4. It appears to be impossible to construct a theory of classical mechanics that is consistent with experiment that does not explicitly, or implicitly, embody absolute space-time in the form of definitions 1 or 2.

## Experimental Determination of an Absolute Reference Frame

Experimentally we have found that space is close to flat although it appears to have enough curvature to form a closed space. It may be flat.

Experimentally we have also found that the observed characteristics of cosmic background radiation (CBR), of the redshift – distance relationship, and of the X-ray background effectively define a preferred local reference frame. The near flatness of space on large distance scales suggests this preferred local reference frame is <u>almost</u> an Absolute Minkowski reference frame. Thus current experiment has found an absolute reference frame – the only question is whether it is local or global. *As Peebles (1993) points out*[50]

*"Blackbody radiation can appear isotropic only in one frame of motion. An observer moving relative to this frame finds that the Doppler shift makes the radiation hotter than average in the direction of motion, cooler in the backward direction. That means CBR acts as an aether, giving a local definition for preferred motion. ... In the standard interpretation, the same preferred comoving rest frame is defined by the CBR, the redshift-distance relation for galaxies, and the X-ray background. ... The evidence*[51] *is that the frames are consistent to perhaps 300 km* $s^{-1}$*."*

*Thus we have an experimental definition of an almost flat, or flat, absolute reference frame.* We conclude with Synge:[52] "The Principle of Equivalence performed the essential office of midwife at the birth of general relativity, but, as Einstein remarked, the infant would never have gotten beyond its longclothes had it not been for

---

49 Eddington (1995) p. 152.

50 Eddington (1995) p. 151-2.

51 M. Aaronson et al, Astrophysical Journal **302**, 536 (1986); R. A. Shafer and A. C. Fabian, in *Early Evolution of the Universe and its Present Structure*, ed. G. O. Abell and G. Chincarini, p. 333 (1983); Rubin (19878).

52 Synge (1960) pp. ix-x.

Minkowski's concept. I suggest that the midwife be now buried with appropriate honours and the facts of an absolute space-time faced."

In the preceding chapters we have defined a unified quantum field theory that embodies the notion of an absolute inertial reference frame (or a set of local preferred inertial reference frames that apply to large locales) in a more direct way than classical general relativity. We defined X coordinates and a Y field in the preferred inertial reference frame of a locale, and then defined two-tier theories that are invariant under special relativistic transformation to other inertial reference frames.

## Curved Space-time Generalization of Two-tier Quantum Gravity

Thus the preceding chapters developed a divergence-free theory of scalar particles and quantum gravity in a flat space-time. In this section we show that a curved space-time version of two-tier quantum field theories including quantum gravity can be developed along the lines pioneered by DeWitt and collaborators. Two-tier curved space-time quantum field theory is based on a mapping from a flat space-time parametrized by y coordinates to a curved space-time parametrized by X coordinates.

The physical picture of the mapping can be visualised using the simple example of a sphere of radius one in three-dimensional space with a coordinate system on the sphere and two planes – one above the sphere and one below it – each with its own flat space coordinate system. Both planes are assumed to be parallel to the disk defined by the crossection of the sphere bounded by the equator of the sphere. A minimum of two coordinate patches are needed to cover a sphere in three dimensions since it necessarily has coordinate singularities.

Let us place a rectangular coordinate system on the top plane. Points on this plane can be mapped onto its northern hemisphere of the sphere in a simple one-to-one fashion. Similarly a rectangular coordinate system can be placed on the bottom plane which can be mapped in a one to one fashion onto the southern hemisphere of the sphere. The top and bottom planes each have a two-dimensional coordinate system that we can choose to be a Cartesian coordinate system in both cases. We will label the coordinates on the top plane $x_t^1$ and $x_t^2$, and the points on the bottom plane as $x_b^1$ and $x_b^2$. Each plane has a flat space metric $g_{tij} = g_{bij} = \delta_{ij}$ for i, j = 1,2 with $\delta_{ij}$ the Kronecker delta.

In addition, just for concreteness, we will place the origin of the top plane coordinate system vertically above the north pole of the sphere, and the origin of the bottom plane coordinate system vertically below the south pole of the sphere.

If we place the sphere at the center of a three dimensional, coordinate system then the points on the sphere (x,y,z) all satisfy:

$$x^2 + y^2 + z^2 = 1 \tag{10.3}$$

We can defined coordinates $u^1$ and $u^2$ for each hemisphere on the surface of the sphere with equations of the form:

$$x_n = f_{1n}(u_n^{\ 1}, u_n^{\ 2}) \tag{10.4}$$
$$y_n = f_{2n}(u_n^{\ 1}, u_n^{\ 2}) \tag{10.5}$$
$$z_n = f_{3n}(u_n^{\ 1}, u_n^{\ 2}) \tag{10.6}$$

for the northern hemisphere, and

$$x_s = f_{1s}(u_s^{\ 1}, u_s^{\ 2}) \tag{10.7}$$
$$y_s = f_{2s}(u_s^{\ 1}, u_s^{\ 2}) \tag{10.8}$$
$$z_s = f_{3s}(u_s^{\ 1}, u_s^{\ 2}) \tag{10.9}$$

for the southern hemisphere.

In addition, we choose $u_n^{\ 1} = u_n^{\ 2} = 0$ at the north pole and $u_s^{\ 1} = u_s^{\ 2} = 0$ at the south pole. The surface of the sphere is curved and each $(u^1, u^2)$ coordinate system has a metric, $g_{nij}$ and $g_{sij}$ for i, j = 1,2 respectively, and a non-zero curvature tensor $R_{nijkl}$ and $R_{sijkl}$.

Now we are allowed to define a simple map of points on the northern hemisphere of the sphere to points on the top plane such as:

$$x_t^{\ 1} = u_n^{\ 1} \tag{10.10}$$
$$x_t^{\ 2} = u_n^{\ 2} \tag{10.11}$$

and of points on the southern hemisphere of the sphere to points on the bottom plane:

$$x_b^{\ 1} = u_s^{\ 1} \tag{10.12}$$
$$x_b^{\ 2} = u_s^{\ 2} \tag{10.13}$$

Thus we can specify the location of events on the sphere on our planes. Note that eqs. 10.4 – 10.9 are *not* a coordinate transformation of the $(u^1, u^2)$ coordinate systems on the sphere and thus the plane can have a different (flat) metric from the sphere.

The preceding example can be simplified by using a cylinder enclosing the sphere instead of two planes. The cylinder, which is a flat surface technically, is aligned so that its axis is parallel to, and cenetered on, the north-south axis of the sphere. Then a map can be made from points on the sphere to points on the cylinder that is similar to a Mercator projection, or from points on the sphere to the cylinder that maps the poles to the ends of the cylinder at + and – infinity.

The preceding discussion shows a clear analogy to our map from the y Minkowski space-time to the curved X space-time using

$$X^{\mu} = y^{\mu} + i\, Y^{\mu}(y)/M_c^{\,2} \tag{10.14}$$

modulo the imaginary term. The y Minkowski space-time has a flat space-time in which we are allowed to choose the Minkowski metric $\eta_{\mu\nu}$. The curved X space-time has an appropriate metric $g_{\mu\nu}(X)$ that can only be transformed to locally inertial coordinates with perhaps a Minkowski metric in the neighborhood of a point. The additional imaginary term does not alter this picture except that the curved X space-time is now a slightly complex manifold in complex space-time.

*Therefore we conclude that our two-tier quantum field theoretic formalism that is erected on eq. 10.14, where the real part of the X space-time was flat, can be extended to curved space-time while maintaining eq. 10.14 if the y space-time consists of coordinate patches analogous to the two planes (or the cylinder) in the example of the sphere. The difference is that we now use a curved space-time background metric $g_{\mu\nu}(X)$ instead of $\eta_{\mu\nu}$ throughout the lagrangian with the exception of $L^Y$ (eq. 9.27).*

In $L^Y$ we continue to use $\eta_{\mu\nu}$ as the metric. As a result $L^Y$ breaks the invariance of the complete lagrangian under general coordinate transformations. Thus an implicit absolute space-time is implied – as it is implicitly in classical General Relativity and in cosmological experiments. This consequence is not disturbing and is physically acceptable for the following reasons:

1. As Bergmann and Synge point out classical general relativity implicitly embodies an absolute space-time.
2. Experiment shows that space in the large (of the order of the Hubble length) is nearly flat although space does appear to be closed. CBR, and other, experimental data suggests that an absolute reference frame exists.

Thus our universe does appear to be in a state of broken general coordinate transformation invariance. Two-tier quantum field theory in curved space-time is not in contradiction with our previous classical general relativistic theories or with our experimental knowledge of the universe. *The full lagrangian theory L is invariant under special relativity. $L - L^Y$ is formally invariant under general coordinate transformations in the X coordinates.*

## Why Are the Y Field Dynamics Independent of the Gravitational Field?

It is evident that the $Y^a$ field is a truly free field in our formulation. In particular, it does not depend on, or interact directly with, the gravitational field as represented by $\sqrt{g}$ and $g_{\mu\nu}$ factors. On the other hand, these quantities depend on the $Y^a$ field through their dependence on the variable $X^{\mu}$.

Thus the role of $Y^a$ is strictly that of a coordinate, and of a field that is parametrized by a set of inertial frame coordinates $y^\mu$. The arguments of Mach supplemented by the arguments of Bergmann and Synge show that a de facto absolute reference frame exists (actually it is the set of inertial reference frames). Therefore we can chose to formulate our theory in an inertial reference frame and require that the theory only be invariant under Lorentz transformations to other inertial reference frames.

In this context it is allowed to have one or more fields like $Y^a$ whose dynamics are not invariant under general coordinate transformations. It it is reasonable to require the particle and gravitational dynamical equations be covariant under general coordinate transformations in X. *Thus a part of the dynamics is invariant under Lorentz transformations – the $Y^a$ sector – but this part of the dynamics is not directly observable; and a part of the dynamics – the observable part – is invariant under general coordinate transformations.*

Some reasons for having a free $Y^a$ field are:

1. It is required to avoid divergences that would appear in perturbation theory if the $Y^a$ were allowed to interact with gravitons. For example an hhYY interaction term causes a divergence to appear by generating a Y particle loop in graviton-graviton scattering.
2. If the $Y^a$ particle interacted with gravity then measurable, classical $Y^a$ fields could be generated in regions with ultra-strong gravitational fields such as the neighborhoods of black holes. In this case we would have new dimensions, allbeit imaginary, for which no experimental evidence currently exists.
3. The Principle of Equivalence has only been shown to apply on the classsical level for real coordinates. Any quantization that uses Minkowskian coordinates, or quasi-Minkowskian coordinates, causes general coordinate transformation invariance to be abandoned ab initio in the quantum regime.

# 11. A Unified Quantum Field Theory of the Known Forces of Nature

## Formulation of Unified Theory

The unification of QED and weak interactions in Electroweak Theory interrelated the theories within the framework of an overall SU(2)⊗U(1) symmetry and thus was significantly more then merely "glueing" the theories together.

The unification of Electroweak theory with Quantum Chromodynamics (QCD) into the Standard Model was a direct combination of these theories in which the symmetries of each respective theory were directly combined: SU(2)⊗U(1) symmetry from Electroweak Theory and SU(3) from QCD to produce the Standard Model with SU(2)⊗U(1)⊗SU(3) symmetry. Electroweak Theory was "glued together" with QCD to produce the Standard Model without an underlying rationale. Nevertheless, the Standard Model is a renormalizable theory that accounts for the vast majority of experimental data.

The present work has two goals: 1.) To make QED, Electroweak Theory, QCD and Quantum Gravity finite and thus remove a major long term defect in Quantum Field Theory, and 2.) To create a unified theory of the known forces of Nature from these pieces. Item 1 has been achieved in the preceding chapters using Two-Tier Quantum Field Theory. This chapter discusses item 2.

In this chapter we propose a finite unified theory of the Standard Model (and any of its variants) and Quantum Gravity within the framework of Two-Tier Quantum Field Theory. As we saw in the case of the Standard Model our unified theory amounts to glueing together the Two-Tier version of the Standard Model with the Two-Tier version of Quantum Gravity. Therefore we regard the theory as provisional in the sense that a deeper unification remains to be formulated.

Nevertheless the Two-Tier Unified Theory may be of some importance beyond the satisfaction of having a finite theory of Nature. It might be a starting point for a deeper, more unified theory. It can be used to address cosmological questions such as the state of the universe immediately after the Big Bang when the size of the universe was of the order of the Planck length or smaller – see Blaha (2004). The interactions in the unified theory become weaker at short distances and thus low order perturbation theory becomes a better approximation to the exact results.

*Unified Two-Tier Standard Model and Quantum Gravity Lagrangian*

We define the Lagrangian, and action, for the two-tier unified quantum field theory of gravitation and the Standard Model as

$$L_{\text{Unified}} = \int d^4y\ \mathscr{L}_{\text{Unified}} \tag{11.1}$$

$$\mathscr{L}_{\text{Unified}} = J\sqrt{g(X)}\ \left(\mathscr{L}_F^{\text{Grav}}(X) + \mathscr{L}_F^{\text{SM}}(X)\right) + \mathscr{L}_C \tag{11.2}$$

with

$$\mathscr{L}_F^{\text{Grav}}(X) = (2\kappa^2)^{-1}{}_X R(X) \tag{11.3}$$

where $\mathscr{L}_F^{\text{SM}}$ is the complete "normal" Quantum Field theory Lagrangian for the Standard Model variant under consideration written in a general coordinate covariant form, g(X) is the absolute value of the determinant of $g_{\mu\nu}$, and J is the Jacobian of eq. A.21. *All particle fields in* $\mathscr{L}_F^{SM}$ *are assumed to be functions of the X coordinate only. The dependence of the particle fields on the "underlying" coordinates* $y^\mu$ *is assumed to be solely through* $X^\mu$. The Lagrangian $\mathscr{L}_{\text{Unified}}$ is a separable Lagrangian of the type of eq. A.26 embodying the composition of extrema described in Appendix A.

As in all cases considered we define the coordinate part of the Lagrangian $\mathscr{L}_C$ as

$$\mathscr{L}_C = -\tfrac{1}{4}\, F_Y^{\mu\nu} F_{Y\mu\nu} \tag{11.4}$$

with

$$F_{Y\mu\nu} = \partial Y_\mu/\partial y^\nu - \partial Y_\nu/\partial y^\mu \tag{11.5}$$

and

$$F_Y^{\mu\nu} = \eta^{\mu\alpha}\eta^{\nu\beta} F_{Y\alpha\beta} \tag{11.6}$$

The development of the physics embodied in the Lagrangian proceeds along the lines described in the previous chapters, except that the gravitational sector must be in the form of a vierbein theory since the full theory contains spin 1/2 particles.

## "Low Energy" Behavior

The low energy sector of the unified theory is defined as the sector with momenta whose values are much less than $M_c$. It is clear from the discussions of the previous chapters that the low energy sector of the unified theory is effectively identical to that of the corresponding conventional quantum field theory if $M_c$ is sufficiently large.

In addition the low energy behavior of the two-tier unified theory in the QED sector closely approximates the results of QED calculations which have been found to agree well with experiment to an extremely high degree of accuracy.

Thus the two-tier unified theory satisfies a Correspondence Principle in the Standard Model sector: The low energy behavior of the two-tier Standard Model sector is the same as the behavior of the conventional Standard Model to a high degree of approximation. In addition the two-tier vierbein Quantum Gravity sector tree diagrams are the same as the conventional vierbein Quantum Gravity tree diagrams to a high degree of accuracy.

## Negative Degree of Divergence – A Finite Unified Theory

The previous discussions of the perturbation theory of matter fields, gauge fields and gravitons show that the theory is finite.

## Unitarity

The unitarity discussions of the various sectors of the unified theory in previous chapters show the unified theory satisfies unitarity. As long as Y excitations are not allowed in in-states they will not appear in out-states. The S matrix is block diagonal and unitary within the physical asymptotic states sector.

# Appendix A. Composition of Extrema in the Calculus of Variations

## A New Paradigm in the Calculus of Variations

The Calculus of Variations has a long and venerable history in Physics and Mathematics. Many problems in Physics and Mathematics have been treated with approaches based on techniques in the Calculus of Variations (see the references at the end of the book). In this book we have developed a unified quantum field theory of the known forces of nature based on a new type of problem, or paradigm, in the Calculus of Variations. One way of viewing the spectrum of problems in the calculus of variations is the following progression.

### *A Classification of Variational Problems*

1. Variational problems in a Euclidean, or Minkowski, flat space such as the minimal distance between two points or the extrema of a field theory Lagrangian.

2. Variational problems seeking extrema on a curved surface such as the shortest distance between points on the surface of a sphere.

The development in this book suggests a third and fourth, possibility, that to the author's knowledge, has not been addressed in the literature:

3. Variational problems where the extrema are determined on a surface that is itself defined as an extremum. The discussions in this book exemplify this pardigm.

4. Variational problems where the extrema are determined on a surface that is itself defined as an extremum that depends on the extrema on the surface. More simply put the extrema, and the surface upon which they are defined, are jointly determined and are interrelated. Fortunately, our unified theory does not use this paradigm. A future theory might.

In the unified theory that we will develop all particle fields including the graviton field are defined as a mapping of a Minkowski space-time y to a "particle" space-time X with the mapping determined as an extremum of a variation of a fundamental field (a type 3 variational problem in the above classification). Our theory could be generalized to include a back-reaction of the particle fields on the fundamental field (a type 4 variational problem in the above classification). We will not discuss this possibility in this book.

## Simple Physical Example – Strings On Springs

In this section we will describe a simple physical example that illustrates a variational problem of type 3 in the Calculus of Variations. We view it as a composition of extrema. (This problem can be addressed using other calculus of Variations techniques.) The approach used in the solution of this problem is similar to the approach used in Two-Tier quantum field theory.

### *A Strings on Springs Mechanics Problem*

Consider a long string or bar that can oscillate (undulate) in a direction perpendicular to its length. Further assume that one end of this bar or string is attached to a spring that cause the entire bar or string to oscillate back and forth in a direction parallel to its long side. This configuration is illustrated in Fig. A.1.

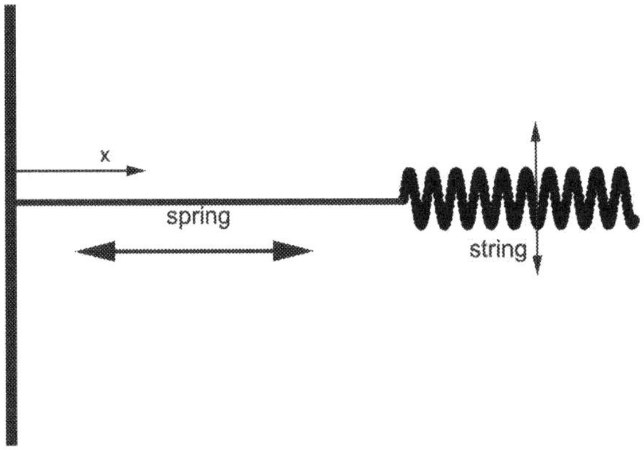


Figure A.1. An oscillating string attached to a spring.

Let x denote the distance to a point on the string when the spring is at equilibrium. If $2\pi$ times the frequency of the spring is $\omega_1$, then the location of this point when the spring is oscillating is

$$X(t) = x + A \sin(\omega_1 t + \phi_1) \tag{A.1}$$

where $\phi_1$ is a phase, and A is the amplitude of the spring oscillation. Then the vertical displacement of a traveling wave on the *string* can take the form

$$\psi(t) = B \sin(\omega_2 t - k_2(x + A \sin(\omega_1 t + \phi_1)) + \phi_2) \qquad (A.2)$$

where B is the amplitude of the string wave, and $k_2$, $\omega_2$ and $\phi_2$ are the parameters of the string wave. These simple mechanical formulae are well known. But they lead to an interesting new application of the ideas of the Calculus of Variations.

Suppose we treat X as an independent variable with X given by eq. (A.1), and with eq. (A.2) written as:

$$\psi(t) = B \sin(\omega_2 t - k_2 X + \phi_2) \qquad (A.3)$$

Defining

$$\psi = \psi(X(t), t) \qquad (A.4)$$

we can specify the dynamics of the above motion by finding the extrema of

$$I = \int \mathscr{L}_\psi \, dX(t) + \int \mathscr{L}_X \, dt \qquad (A.5)$$

where the Lagrangian terms are

$$\mathscr{L}_\psi = \tfrac{1}{2} \{ \mu (\partial\psi/\partial t)^2 - Y (\partial\psi/\partial X)^2 \} \qquad (A.6)$$

with $\mu$ and Y being constants, and

$$\mathscr{L}_X = \tfrac{1}{2} \{ m(\partial X/\partial t)^2 - k(X - x)^2 \} \qquad (A.7)$$

where m and k are constants, and where x is a parameter. Applying Hamilton's Principle, and performing independent variations of X and $\psi$ yields the Lagrangian equations:

$$\frac{\partial \mathscr{L}_\psi}{\partial \psi} - \frac{\partial}{\partial X} \frac{\partial \mathscr{L}_\psi}{\partial(\partial\psi/\partial X)} - \frac{\partial}{\partial t} \frac{\partial \mathscr{L}_\psi}{\partial(\partial\psi/\partial t)} = 0 \qquad (A.8)$$

and

$$\frac{\partial \mathscr{L}_X}{\partial X} - \frac{\partial}{\partial t}\frac{\partial \mathscr{L}_X}{\partial(\partial X/\partial t)} = 0 \tag{A.9}$$

The resulting equations of motion are:

$$\mu\, \partial^2\psi/\partial t^2 - Y\, \partial^2\psi/\partial X^2 = 0 \tag{A.10}$$

and

$$m\, \partial^2 X/\partial t^2 + k(X - x) = 0 \tag{A.11}$$

with the solutions given in eqs. A.1 and A.2.

The procedure that we use to obtain these results may look a bit strange but they illustrate a type 3 problem in the Calculus of Variations involving the composition of extrema—the composition of an extremum that specifies a manifold in a space (possibly including all of space in a $R^n \rightarrow R^n$ mapping) with an extremum determining a function on that manifold. The procedure is described in detail in the next section.

## The Composition of Extrema – Lagrangian Formulation

In this section we will explore the general case of the composition of extrema for fields. We will discuss the case of a scalar field $\phi$ that is a function of a vector field $X^\mu$ in a D-dimensional space with coordinate variables that we will denote as $y^\mu$. (The discussion for other types of fields is a straightforward extension of this discussion.) Thus

$$\phi = \phi(X) \tag{A.12}$$

and

$$X^\mu = X^\mu(y) \tag{A.13}$$

We assume that the dynamics can be described by a Lagrangian formulation using an extension of Hamilton's principle:

$$I = \int \mathscr{L} d^4y \tag{A.14}$$

with

$$\mathscr{L} = \mathscr{L}(\phi(X), \partial\phi/\partial X^\nu, X^\mu(y), \partial X^\mu(y)/\partial y^\nu, y) \tag{A.15}$$

If we perform a standard variation[53] in $\phi$ for fixed y (and thus fixed X) we find

$$\delta \mathrm{I} = \int [\delta\phi\, \partial\mathscr{L}/\partial\phi + \delta(\partial\phi/\partial \mathrm{X}^{\nu})\, \partial\mathscr{L}/\partial(\partial\phi/\partial \mathrm{X}^{\nu})]\, d^4y \qquad (A.16)$$

We can rewrite the variation in the derivative of $\phi$ as

$$\delta(\partial\phi/\partial \mathrm{X}^{\nu}) = \partial(\delta\phi)/\partial \mathrm{X}^{\nu} \qquad (A.17)$$

$$= \partial y^{\mu}/\partial \mathrm{X}^{\nu}\, \partial(\delta\phi)/\partial y^{\mu} \qquad (A.18)$$

with an implied summation over repeated indices. After substituting eq. A.18 in eq. A.16, and performing an integration by parts (and discarding the surface term which is assumed to yield zero in the standard fashion) we obtain:

$$\delta \mathrm{I} = \int \delta\phi\, \{\partial\mathscr{L}/\partial\phi - \partial/\partial y^{\mu}\, [\partial\mathscr{L}/\partial(\partial\phi/\partial \mathrm{X}^{\nu})\, \partial y^{\mu}/\partial \mathrm{X}^{\nu}\,]\}\, d^4y$$

Since the variation of $\delta\phi$ is arbitrary we conclude

$$\partial\mathscr{L}/\partial\phi - \partial/\partial y^{\mu}\, [\partial\mathscr{L}/\partial(\partial\phi/\partial \mathrm{X}^{\nu})\, \partial y^{\mu}/\partial \mathrm{X}^{\nu}] = 0 \qquad (A.19a)$$

The second term in eq. A.19a shows the effect of the dependence of $\phi$ on the field X, $\phi = \phi(\mathrm{X})$, rather than directly on the coordinate system y.

Similarly we can perform a variation in $\mathrm{X}^{\mu}$ and obtain

$$\partial\mathscr{L}/\partial \mathrm{X}^{\mu} - \partial/\partial y^{\nu}\, [\partial\mathscr{L}/\partial(\partial \mathrm{X}^{\mu}/\partial y^{\nu})] = 0 \qquad (A.19b)$$

The X field defines a "manifold" or, more properly, specifies a transformation from $R^n \rightarrow R^n$. If we make standard assumptions about the mapping: that it is continuous and piece-wise invertible, then we can establish the following lemmas:

**Lemma 1**: *If the transformation $X^{\mu} = X^{\mu}(y)$ is a transformation from $R^n \rightarrow R^n$ that is of class $C'$ and piece-wise invertible, then*

53 Bogoliubov, N. N., & Shirkov, D. V., Volkoff, G. M. (tr), *Introduction to the Theory of Quantized Fields* (Wiley-Interscience, New York, 1959); Goldstein H., *Classical Mechanics* (Addison-Wesley, Reading, MA 1965).

$$\frac{\partial}{\partial y^\nu}\frac{\partial y^\nu}{\partial X^\mu} = -\frac{\partial \ln J}{\partial X^\mu} \tag{A.20}$$

*where*

$$J = |\partial(X)/\partial(y)| \tag{A.21}$$

*is the absolute value of the Jacobian of the transformation.*

**Proof:**
Consider two equivalent forms of an integral:

$$I = \int \mathscr{L} J\, d^4y = \int \mathscr{L}\, d^4X$$

where $\mathscr{L}$ is specified as in eq. A.15. Then the first expression for I leads to eq. A.19a which can be written in the form

$$\partial\mathscr{L}/\partial\phi - \partial/\partial X^\mu\, [\partial\mathscr{L}/\partial(\partial\phi/\partial X^\mu)] - \partial\mathscr{L}/\partial(\partial\phi/\partial X^\mu)\{\partial[J\partial y^\nu/\partial X^\mu]/\partial y^\nu\} = 0$$

Using the second expression for I above we obtain the following equation by variation in $\phi$:

$$\partial\mathscr{L}/\partial\phi - \partial/\partial X^\mu\, [\partial\mathscr{L}/\partial(\partial\phi/\partial X^\mu)] = 0$$

Comparing these two expressions and realizing that $\partial[J\partial y^\nu/\partial X^\mu]/\partial y^\nu$ is totally independent of $\phi$ and its derivatives leads us to conclude

$$\partial[J\partial y^\nu/\partial X^\mu]/\partial y^\nu = 0 \tag{A.22}$$

It is a general relationship for a transformation between X and y based on continuity and piece-wise invertibility. After a few elementary manipulations eq. A.22 can be rewritten in the form of eq. A.20. ■

**Lemma 2**: *If the transformation $X^\mu = X^\mu(y)$ is a transformation from $R^n \rightarrow R^n$ that is of class C′ and piece-wise invertible and* $\mathscr{L} = \mathscr{L}(\phi(X),\ \partial\phi/\partial X^\nu,\ X^\mu(y),\ \partial X^\mu(y)/\partial y^\nu,\ y)$, *then*

$$\partial\mathscr{L}/\partial(\partial\phi/\partial X^{\nu})\,\partial y^{\mu}/\partial X^{\nu} = \partial\mathscr{L}/\partial(\partial\phi/\partial y^{\mu}) \qquad \text{(A.23)}$$

**Proof:**

Let us express $\mathscr{L}$ as a power series in derivatives of $\phi$:

$$\mathscr{L} = \sum_{n=0} a_{n\mu_1\mu_2\ldots\mu_n}(\phi(X), X^{\mu}(y), \partial X^{\mu}(y)/\partial y^{\nu}, y) \prod_{j=1}^{n} \partial\phi/\partial X^{\mu_j}$$

which can rewritten using piece-wise invertibility as

$$\mathscr{L} = \sum_{n=0} a_{n\mu_1\mu_2\ldots\mu_n}(\phi(X), X^{\mu}(y), \partial X^{\mu}(y)/\partial y^{\nu}, y) \prod_{j=1}^{n} \partial\phi/\partial y^{\nu_j}\,\partial y^{\nu_j}/\partial X^{\mu_j}$$

Taking the derivative of this equation with respect to $\partial\phi/\partial y^{\mu}$ immediately yields the result. ■

Eq. A.23 enables us to rewrite eq. A.19a as:

$$\partial\mathscr{L}/\partial\phi - \partial/\partial y^{\mu}\,[\partial\mathscr{L}/\partial(\partial\phi/\partial y^{\mu})] = 0 \qquad \text{(A.24)}$$

which is as one would expect.

In order to get a feeling for the effect of eq. A.19a we will look at a simple example where we specify the relation of the X and y variables directly. Then we will look at the composition of extrema where the transformation between X and y is itself determined as an extremum solution.

*Example: a hyperplane*

We assume eq. A.19b yields the transformation:

$$X^{i} = ay^{i} \qquad \text{for } i = 1,2,3$$

$$X^{0} = 0$$

Then eq. A.19a becomes

$$\partial\mathcal{L}/\partial\phi - \partial/\partial y^i\,[\partial\mathcal{L}/\partial(\partial\phi/\partial y^i)] = 0 \qquad \text{(A.25)}$$

with the time derivative disappearing. Effectively the variation of $\phi$ on the hyperplane $X^0 = 0$ is determined by the differential equation generated by A.25. On this hyperplane the transformation between the X and y variables is invertible.

*Coordinate Transformation Determined as an Extremum Solution*

We now develop a formalism that determines a mapping from space onto itself as the solution of an extremum problem and also determines the dynamics of one or more fields as a function of this mapping. To this author's knowledge this area in the Calculus of Variations – the determination of an extremum on a manifold where the manifold itself is determined by an extremum – has not been previously explored. We will also develop a hamiltonian formulation. Then we will proceed to quantize the theory.

*Separable Lagrangian Case*

Although there are many forms that the composition of extrema could take, one fairly general form that is directly useful in quantum field theory applications is based on a Lagrangian that can be split into two parts which we will call a *separable Lagrangian*:

$$\mathcal{L} = \mathcal{L}_F\, J + \mathcal{L}_C(X^\mu(y), \partial X^\mu(y)/\partial y^\nu, y) \qquad \text{(A.26)}$$

where J is defined in eq. A.21, where $\mathcal{L}_F$ contains all the dynamics of the fields and their interactions, and where $\mathcal{L}_C$ defines the coordinate mapping as an extremum solution. The procedure to determine the differential equations that specify the mapping, and the field equations that specify field interactions and evolution, is to vary in the coordinates $X^\mu$ and in the fields independently, using Hamilton's Principle. The extrema are to be determined for

$$I = \int \mathcal{L}\, d^4y \qquad \text{(A.27)}$$

We will begin by considering the case of one scalar field:

$$\mathcal{L}_F = \mathcal{L}_F(\phi(X), \partial\phi/\partial X^\nu) \qquad \text{(A.28)}$$

and

$$\mathcal{L}_C = \mathcal{L}_C(X^\mu(y), \partial X^\mu(y)/\partial y^\nu, y) \qquad \text{(A.29)}$$

Eq. A.27 can be written in the form:

$$I = \int \mathscr{L}_F(\phi(X), \partial\phi/\partial X^\nu)\, dX + \int \mathscr{L}_C(X^\mu(y), \partial X^\mu(y)/\partial y^\nu, y)\, d^4y \qquad \text{(A.30)}$$

using the Jacobian to transform to an integral over dX in the first term. A standard variation of $\phi$ and the application of Hamilton's Principle yields

$$\partial\mathscr{L}_F/\partial\phi - \partial/\partial X^\mu\, [\partial\mathscr{L}_F/\partial(\partial\phi/\partial X^\mu)] = 0 \qquad \text{(A.31)}$$

reflecting the fact that $\phi$ is a function of $X^\mu$ only, with $X^\mu$ a function of the y coordinates.

Next we perform a variation of $X^\mu$ determining the mapping from $y \rightarrow X$ as an extremum of the integral in eq. A.27. We note the piece-wise invertibility of the coordinate mapping $X^\mu(y)$ allows us to write the Jacobian J as a function of $y^\mu$ only. A standard variation of $X^\mu$ and the application of Hamilton's Principle yields

$$\partial\mathscr{L}_C/\partial X^\mu - \partial/\partial y^\nu\, [\partial\mathscr{L}_C/\partial(\partial X^\mu/\partial y^\nu)] = 0 \qquad \text{(A.32)}$$

*Klein-Gordon Example*

The Klein-Gordon scalar field theory furnishes us with a simple example of the application of the preceding development. The Lagrangian is

$$\mathscr{L}_F = \tfrac{1}{2}\, [\, (\partial\phi/\partial X^\nu)^2 - m^2\phi^2\,] \qquad \text{(A.33)}$$

From eq. A.31 we obtain the field equation:

$$(\Box + m^2)\, \phi(X) = 0 \qquad \text{(A.34)}$$

where

$$\Box = \partial/\partial X^\nu\, \partial/\partial X_\nu \qquad \text{(A.34a)}$$

A fourier representation of the solution of eq. A.34 is:

$$\phi(X) = \int dp\, \delta(p^2 - m^2)\theta(p^0)\, [A(p)\, e^{-ip\cdot X} + A(p)^*\, e^{ip\cdot X}] \qquad \text{(A.35)}$$

where A(k) is a function of k and * indicates complex conjugation.

The determination of $X^\mu(y)$ depends on the Lagrangian $\mathscr{L}_C$ and the solutions of eq. A.3A. If we chose

$$\mathscr{L}_C = -\tfrac{1}{2}\,(\partial X^\mu/\partial y^\nu)^2 \tag{A.36}$$

Then we obtain the equation

$$\Box\, X^\mu = 0 \tag{A.37}$$

with the solution

$$X^\mu = \int dk\,\delta(k^2)\theta(k^0)\,[a^\mu(k)\,e^{-ik\cdot y} + a^\mu(k)^*\,e^{ik\cdot y}] \tag{A.38}$$

where $a^\mu(k)$ are complex vector functions of k in general. (We ignore positivity issues for the moment.) Substitution of eq. A.38 in eq. A.35 yields an expression with a form reminiscent of bosonic string expressions.[54] We will take up this point later in subsequent chapters.

## The Composition of Extrema – Hamiltonian Formulation

The previous section established a Lagrangian formulation of dynamics based on the composition of extrema. In this section we will develop an equivalent hamiltonian formulation. We will assume a Minkowskian space-time with $X^0$ and $y^0$ playing the role of the time coordinates in the respective coordinate systems.

Initially, we will assume a scalar field $\phi$ with a Lagrangian of the form in eq. A.15 and define canonical momenta with

$$\Pi_\phi = \partial\mathscr{L}/\partial\dot{\phi} \equiv \partial\mathscr{L}/\partial(\partial\phi/\partial X^\mu)\,\partial y^0/\partial X^\mu \tag{A.39}$$

$$\Pi_X{}^\mu = \partial\mathscr{L}/\partial\dot{X}_\mu \tag{A.40}$$

where

$$\dot{\phi} = \partial\phi/\partial y^0 \equiv \partial\phi/\partial X^\mu\,\partial X^\mu/\partial y^0 \tag{A.41}$$

$$\dot{X}^\mu = \partial X^\mu/\partial y^0 \tag{A.42}$$

Then we define the hamiltonian density as

$$\mathscr{H} = \Pi_\phi\,\dot{\phi} + \Pi_X{}^\mu\,\dot{X}_\mu - \mathscr{L}(\phi(X), \partial\phi/\partial X^\nu, X^\mu(y), \partial X^\mu(y)/\partial y^\nu, y) \tag{A.43}$$

and the hamiltonian

---

54 See for example Polchinski (1998) and Bailin (1994).

$$H = \int \mathcal{H} \, d^3y \tag{A.44}$$

The hamiltonian density has the general form

$$\mathcal{H} = \mathcal{H}(\phi(X), \partial\phi/\partial X^i, \Pi_\phi, X^\mu(y), \partial X^\mu(y)/\partial y^j, \Pi_X{}^\mu, y^\nu) \tag{A.45}$$

for the case of one scalar field where the indices i and j represent space coordinates; time coordinates are assigned index value 0.

If we calculate the differential change in H using eq. A.45 we obtain

$$dH = \int \{\partial\mathcal{H}/\partial\phi \, d\phi + \partial\mathcal{H}/\partial\Pi_\phi \, d\Pi_\phi - \partial/\partial y^\nu[\partial\mathcal{H}/\partial(\partial\phi/\partial X^i)\partial y^\nu/\partial X^i]d\phi +$$

$$+ \partial\mathcal{H}/\partial X^\mu \, dX^\mu + \partial\mathcal{H}/\partial\Pi_X{}^\mu \, d\Pi_X{}^\mu - \partial/\partial y^j \, [\partial\mathcal{H}/\partial(\partial X^\mu/\partial y^j)] \, dX^\mu\} \, d^3y \tag{A.46}$$

after some partial integrations. (Repeated indices indicate summations. Indices labeled i and j indicate space coordinates. Greek indices include all space-time components of a variable.)

Expressing the differential in H using eq. A.43 we obtain

$$dH = \int dy \, \{\Pi_\phi \, d\dot{\phi} + \dot{\phi} \, d\Pi_\phi - \partial\mathcal{L}/\partial\phi \, d\phi - \partial\mathcal{L}/\partial(\partial\phi/\partial X^\mu)d(\partial\phi/\partial X^\mu) +$$

$$+ \Pi_X{}^\mu \, d\dot{X}^\mu + \dot{X}^\mu \, d\Pi_X{}^\mu - \partial\mathcal{L}/\partial X^\mu \, dX^\mu - \partial\mathcal{L}/\partial(\partial X^\mu/\partial y^j)d(\partial X^\mu/\partial y^j)\} \tag{A.47a}$$

After some manipulations we find

$$dH = \int \{ \dot{\phi} \, d\Pi_\phi + \dot{X}_\mu d\Pi_X{}^\mu - \partial/\partial y^0 \, \Pi_\phi \, d\phi - \partial/\partial y^0 \, \Pi_X{}^\mu \, dX_\mu\} \, dy \tag{A.47b}$$

using the equations of motion eqs. A.19a and A.19b.

Comparing eqs A.46 and A.47 we obtain Hamilton's equations in the case of the composition of extrema:

$$\dot{\phi} = \partial\mathcal{H}/\partial\Pi_\phi \tag{A.48a}$$

$$\dot{\Pi}_\phi = -\partial\mathcal{H}/\partial\phi + \partial/\partial y^\nu\, [\partial\mathcal{H}/\partial(\partial\phi/\partial X^j)\; \partial y^\nu/\partial X^j] \tag{A.48b}$$

$$\dot{X}_\mu = \partial\mathcal{H}/\partial\Pi_X{}^\mu \tag{A.48c}$$

$$\dot{\Pi}_X{}^\mu = -\partial\mathcal{H}/\partial X^\mu + \partial/\partial y^j\, [\partial\mathcal{H}/\partial(\partial X^\mu/\partial y^j)] \tag{A.48d}$$

where

$$\dot{\Pi}_\phi = \partial\, \Pi_\phi/\partial y^0 \tag{A.49a}$$

$$\dot{\Pi}_X{}^\mu = \partial\Pi_X{}^\mu/\partial y^0 \tag{A.49b}$$

## Translational Invariance

If the Lagrangian of a field theory has no explicit dependence on the coordinates then one expects translational invariance accompanied by a conservation law for an energy-momentum stress tensor. We will show this is the case for Lagrangians implementing the composition of extrema. We assume a Lagrangian without an explicit dependence on the coordinates $y^\nu$:

$$\mathcal{L} = \mathcal{L}(\phi(X), \partial\phi/\partial X^\nu, X^\mu(y), \partial X^\mu(y)/\partial y^\nu) \tag{A.50}$$

Under an infinitesimal displacement,

$$y'^\nu = y^\nu + \epsilon^\nu \tag{A.51a}$$

$$\delta\phi = \phi(X(y + \epsilon)) - \phi(X(y))$$

$$= \epsilon^a\, \partial\phi/\partial y^a \tag{A.51b}$$

$$\delta X^\mu = \epsilon^a\, \partial X^\mu/\partial y^a \tag{A.51c}$$

$$\delta(\partial\phi/\partial X^\mu) = \epsilon^a\, \partial(\partial\phi/\partial y^a)/\partial X^\mu \tag{A.51d}$$

$$\delta(\partial X^\mu/\partial y^\nu) = \in^a \partial(\partial X^\mu/\partial y^a)/\partial y^\nu \tag{A.51e}$$

and the Lagrangian changes by

$$\delta\mathscr{L} = \in^a \partial\mathscr{L}/\partial y^a \tag{A.52}$$

The change can also be expressed in terms of the changes in the fields, their derivatives and the mapping $X^\mu$:

$$\delta\mathscr{L} = \partial\mathscr{L}/\partial\phi\,\delta\phi + \partial\mathscr{L}/\partial(\partial\phi/\partial X^\mu)\,\delta(\partial\phi/\partial X^\mu) + \partial\mathscr{L}/\partial X^\mu\,\delta X^\mu + + \partial\mathscr{L}/\partial(\partial X^\mu/\partial y^\nu)\,\delta(\partial X^\mu/\partial y^\nu) \tag{A.53}$$

Combining eqs. A.51, A.52 and A.53 we obtain (after some manipulations):

$$\in^\nu \partial/\partial y_\mu\,\mathscr{T}_{\mu\nu} = 0 \tag{A.54}$$

where

$$\mathscr{T}_{\mu\nu} = -g_{\mu\nu}\mathscr{L} + \partial\mathscr{L}/\partial(\partial\phi/\partial X^\delta)\,\partial y_\mu/\partial X^\delta\,\partial\phi/\partial y^\nu + \partial\mathscr{L}/\partial(\partial X^\delta/\partial y_\mu)\partial X^\delta/\partial y^\nu \tag{A.55a}$$

or, alternately using Lemma 2,

$$\mathscr{T}_{\mu\nu} = -\,g_{\mu\nu}\mathscr{L} + \partial\mathscr{L}/\partial(\partial\phi/\partial y_\mu)\,\partial\phi/\partial y^\nu + \partial\mathscr{L}/\partial(\partial X^\delta/\partial y_\mu)\,\partial X^\delta/\partial y^\nu \tag{A.55b}$$

Since $\in^a$ is an arbitrary displacement we obtain the conservation law:

$$\partial/\partial y_\mu\,\mathscr{T}_{\mu\nu} = 0 \tag{A.56}$$

Eq. A.56 implies the energy-momentum vector

$$P_\beta = \int d^3y\,\mathscr{T}_{0\beta} \tag{A.57}$$

is conserved. We note

$$\partial/\partial y^0\, P_\beta = 0 \qquad \text{(A.58)}$$

since eq. A.56 and A.57 can be used to obtain the integral of a divergence, which results in zero.

The hamiltonian (eqs. A.43-44) is

$$H = P_0 \qquad \text{(A.59)}$$

We note for later use that the total energy, H, which is conserved, contains a term that represents the energy in the $X^\mu$ mapping. Thus energy can be exchanged in principle between the $\phi$ field sector and the $X^\mu$ sector.

## Lorentz Invariance and Angular Momentum Conservation

We can also verify Lorentz invariance and obtain the form of the conserved angular momentum by considering the effect of an infinitesimal Lorentz transformation. We will consider the case of a scalar field $\phi$.

Under an infinitesimal Lorentz transformation ($\epsilon_{\mu\nu} = -\epsilon_{\nu\mu}$):

$$y'_\mu = y_\mu + \delta y_\mu = y_\mu + \epsilon_{\mu\nu} y^\nu \qquad \text{(A.60a)}$$

$$\delta\phi = \phi(X(y')) - \phi(X(y))$$

$$= \epsilon^{\mu\nu}\, y_\nu\, \partial\phi/\partial X^a\, \partial X^a/\partial y^\mu \qquad \text{(A.60b)}$$

$$\delta X^\mu = S^\mu{}_a X^a(y') - X^\mu(y) \qquad \text{(A.60c)}$$

$$= \epsilon^\mu{}_a X^a(y) + \partial X^\mu/\partial y^\beta\, \delta y^\beta \qquad \text{(A.60d)}$$

where $S^\mu{}_a$ is the matrix for the Lorentz transformation of a vector. (If $X^\mu$ were a gauge field then an additional operator gauge term would have to be added to eq. A.60d.)

The Lagrangian changes by

$$\delta\mathscr{L} = \epsilon^{\mu\nu}\, y_\nu\, \partial\mathscr{L}/\partial y^\mu \qquad \text{(A.61)}$$

under the infinitesimal Lorentz transformation. The change in the Lagrangian can also be expressed as:

$$\delta\mathscr{L} = \partial\mathscr{L}/\partial\phi\,\delta\phi + \partial\mathscr{L}/\partial(\partial\phi/\partial X^\mu)\,\delta(\partial\phi/\partial X^\mu) + \partial\mathscr{L}/\partial X^\mu\,\delta X^\mu + \\ + \partial\mathscr{L}/\partial(\partial X^\mu/\partial y^\nu)\,\delta(\partial X^\mu/\partial y^\nu) \qquad \text{(A.62)}$$

Combining eqs. A.61 and A.62, and substituting and simplifying terms leads to:

$$\in_{\mu\nu} \partial/\partial y^\sigma \mathscr{M}^{\sigma\mu\nu} = 0 \qquad \text{(A.63)}$$

where

$$\mathscr{M}^{\sigma\mu\nu} = (g^{\mu\sigma}y^\nu - g^{\nu\sigma}y^\mu)\mathscr{L} + \partial\mathscr{L}/\partial(\partial\phi/\partial X^\alpha)\,\partial y^\sigma/\partial X^\alpha\,(y^\mu\partial\phi/\partial y_\nu - y^\nu\partial\phi/\partial y_\mu) + \\ + \partial\mathscr{L}/\partial(\partial X^\delta/\partial y^\sigma)\,(g^{\delta\nu}X^\mu - g^{\delta\mu}X^\nu + y^\mu\,\partial X^\delta/\partial y_\nu - y^\nu\,\partial X^\delta/\partial y_\mu) \qquad \text{(A.64)}$$

The conserved angular momentum is:

$$M^{\mu\nu} = \int d^3y\,\mathscr{M}^{0\mu\nu} \qquad \text{(A.65)}$$

with

$$\partial M^{\mu\nu}/\partial y^0 = 0 \qquad \text{(A.66)}$$

The angular momentum density can be written in the familiar form:

$$\mathscr{M}^{\sigma\mu\nu} = y^\mu\,\mathscr{T}^{\sigma\nu} - y^\nu\,\mathscr{T}^{\sigma\mu} + \partial\mathscr{L}/\partial(\partial X^\delta/\partial y^\sigma)\,(g^{\delta\nu}X^\mu - g^{\delta\mu}X^\nu) \qquad \text{(A.67)}$$

taking account of the vector nature of $X^\mu$. The spatial part of $M^{\mu\nu}$ is the angular momentum.

## Internal Symmetries

We will now consider the case of a set of scalar fields $\phi_r$ in a Lagrangian with an internal symmetry. Under a local transformation

$$\phi_r(X) \to \phi_r(X) - i\in\lambda_{rs}\,\phi_s(X) \qquad \text{(A.68)}$$

If the Lagrangian is invariant under this transformation, then

$$\delta\mathscr{L} = 0 = \partial\mathscr{L}/\partial\phi_r\delta\phi_r + \partial\mathscr{L}/\partial(\partial\phi_r/\partial X^\alpha)\,\delta(\partial\phi_r/\partial X^\alpha) \qquad \text{(A.69)}$$

Using the equation of motion eq. A.19a satisfied by all the components $\phi_r$ we obtain a conserved current:

$$\mathscr{J}^{\nu} = -i\, \partial\mathscr{L}/\partial(\partial\phi_r/\partial X^{\delta})\, \partial y^{\nu}/\partial X^{\delta}\, \lambda_{rs}\, \phi_s \tag{A.70}$$

which satisfies

$$\partial\mathscr{J}^{\nu}/\partial y^{\nu} = 0 \tag{A.71}$$

The conserved charge is

$$Q = \int d^3y\, \mathscr{J}^0 \tag{A.72}$$

$$\partial Q/\partial y^0 = 0 \tag{A.73}$$

## Separable Lagrangians

We now consider the case of a separable Lagrangian such as in eq. A.26. Adopting the definitions:

$$\phi' = \partial\phi/\partial X^0 \tag{A.74}$$

$$X_{\mu}' = \partial X_{\mu}/\partial y^0 \tag{A.75}$$

we define canonical momenta as

$$\pi_{\phi} = \partial\mathscr{L}/\partial\phi' \equiv \partial\mathscr{L}/\partial(\partial\phi/\partial X^0) \tag{A.76}$$

$$\pi_X^{\ \mu} = \partial\mathscr{L}/\partial X_{\mu}' \equiv \partial\mathscr{L}/\partial(\partial X_{\mu}/\partial y^0) \tag{A.77}$$

We now define the separable hamiltonian density as

$$\mathscr{H}_s = J\pi_{\phi}\,\phi' + \pi_X^{\ \mu}\, X_{\mu}' - \mathscr{L}_s \tag{A.78}$$

where J is the Jacobian (eq. A.21) and

$$H_s = \int \mathscr{H}_s\, d^3y \tag{A.79}$$

The separable Lagrangian (from eq. A.26) is:

$$\mathscr{L}_s = \mathscr{L}_F(\phi(X), \partial\phi/\partial X^\mu)\, J + \mathscr{L}_C(X^\mu(y), \partial X^\mu(y)/\partial y^\nu, y) \qquad (A.80)$$

In the case of one scalar field the separable hamiltonian density has the general form

$$\mathscr{H}_s = \mathscr{H}_S(\phi(X), \pi_\phi, \partial\phi/\partial X^i, X^\mu(y), \pi_X{}^\mu, \partial X^\mu(y)/\partial y^j, y^\nu) \qquad (A.81)$$

where the indices i and j indicate spatial components. In particular, the terms in the separable hamiltonian are:

$$\mathscr{H}_s = \mathscr{H}_F\, J + \mathscr{H}_C \qquad (A.82)$$

with

$$\mathscr{H}_F(\phi(X), \pi_\phi, \partial\phi/\partial X^i) = \pi_\phi\, \phi' - \mathscr{L}_F \qquad (A.83)$$

$$\mathscr{H}_C(X^\mu(y), \pi_X{}^\mu, \partial X^\mu(y)/\partial y^j, y^\nu) = \pi_X{}^\mu\, X_\mu{}' - \mathscr{L}_C \qquad (A.84)$$

where J is the absolute value of the Jacobian defined in A.21.

We now define the time integral of H as we did in eq. A.14 when considering the Lagrangian formulation:

$$G = \int dy^0\, H_s \qquad (A.85)$$

Thus G is an integral over all space-time coordinates. Using G we can develop a hamiltonian formulation. First we calculate the differential change in G. Using eqs. A.81-2 and A.85 we obtain

$$\begin{aligned} dG = \int \{ & J\, \partial\mathscr{H}_F/\partial\phi\, d\phi + J\, \partial\mathscr{H}_F/\partial\pi_\phi\, d\pi_\phi + \\ & + J\, \partial\mathscr{H}_F/\partial(\partial\phi/\partial X^j)\, d(\partial\phi/\partial X^j) + \partial\mathscr{H}_C/\partial X^\mu\, dX^\mu + \\ & + \partial\mathscr{H}_C/\partial\pi_X{}^\mu\, d\pi_X{}^\mu + \partial\mathscr{H}_C/\partial(\partial X^\mu/\partial y^j)\, d(\partial X^\mu/\partial y^j) \}\, d^4y \end{aligned} \qquad (A.86)$$

with summations implied by repeated indices. (Index labels i and j label spatial coordinates only; Greek indices label space-time coordinates.) Rewriting dG as two integrals and performing partial integrations yields:

$$dG = \int d^4X \{ \partial\mathcal{H}_F / \partial\phi \, d\phi + \partial\mathcal{H}_F / \partial\pi_\phi d\pi_\phi - \partial/\partial X^j [\partial\mathcal{H}_F / \partial(\partial\phi/\partial X^j)] \, d\phi \} +$$
$$+ \int d^4y \{ \partial\mathcal{H}_C / \partial X^\mu \, dX^\mu + \partial\mathcal{H}_C / \partial\pi_X{}^\mu \, d\pi_X{}^\mu - \partial/\partial y^j [\partial\mathcal{H}_C / \partial(\partial X^\mu/\partial y^j)] \, dX^\mu \} \quad (A.87)$$

*Alternately, expressing the differential in G using eqs. A.82-4 we obtain*

$$dG = \int d^4X \{ \pi_\phi \, d\phi' + \phi' d\pi_\phi - \partial\mathcal{L}_F / \partial\phi \, d\phi - \partial\mathcal{L}_F / \partial(\partial\phi/\partial X^\mu) d(\partial\phi/\partial X^\mu) \} +$$
$$+ \int d^4y \{ \pi_{X\mu} \, dX^{\mu\prime} + X^{\mu\prime} \, d\pi_{X\mu} - \partial\mathcal{L}_C / \partial X^\mu \, dX^\mu - \partial\mathcal{L}_C / \partial(\partial X^\mu/\partial y^j) d(\partial X^\mu/\partial y^j) \} \quad (A.88)$$

which becomes

$$dG = \int d^4X \{ -\pi_\phi{}' \, d\phi + \phi' \, d\pi_\phi \} + \int d^4y \{ -\pi_{X\mu}{}' \, dX^\mu + X^{\mu\prime} \, d\pi_{X\mu} \} \quad (A.89)$$

using the equations of motion eqs. A.31-2.

Comparing eqs A.87 and A.89 we obtain Hamilton's equations for the case of the composition of extrema for a separable Lagrangian:

$$\phi' = \partial\mathcal{H}_F / \partial\pi_\phi \quad (A.90)$$

$$\pi_\phi{}' = -\partial\mathcal{H}_F / \partial\phi + \partial/\partial X^j \, [\partial\mathcal{H}_F / \partial(\partial\phi/\partial X^j)] \quad (A.91)$$

$$X_\mu{}' = \partial\mathcal{H}_C / \partial\pi_X{}^\mu \quad (A.92)$$

$$\pi_{X\mu}{}' = -\partial\mathcal{H}_C / \partial X^\mu + \partial/\partial y^j \, [\partial\mathcal{H}_C / \partial(\partial X^\mu/\partial y^j)] \quad (A.93)$$

where

$$\pi_\phi{}' = \partial \, \pi_\phi / \partial X^0 \quad (A.94)$$

$$\pi_{X\mu}{}' = \partial \, \pi_{X\mu} / \partial X^0 \quad (A.95)$$

*Notice that* $\mathscr{L}_F$, $\mathscr{H}_F$ *and* $\pi_\phi$ *have precisely the same form, as a function of* $X^\mu$, *as one sees in a conventional field theory formalism.* Yet $X^\mu$ is a mapping/function of the coordinates y. In reality, it can be viewed as a field as we shall see.

## Separable Lagrangians and Translational Invariance

The general rule for conventional Lagrangians is: if a Lagrangian has no explicit dependence on the coordinates then translational invariance follows accompanied by a conservation law for an energy-momentum tensor. We will show that this rule needs modification for separable Lagrangians that implement the composition of extrema.

Consider the Lagrangian:

$$\mathscr{L}_S = J\ \mathscr{L}_F(\phi(X), \partial\phi/\partial X^\mu) + \mathscr{L}_C(X^\mu(y), \partial X^\mu(y)/\partial y^\nu) \tag{A.96}$$

in which the $X^\mu$ play a dual role as both fields and coordinates. Let us consider a variation in $X^\mu$:

$$X^\mu(y) \rightarrow X^\mu(y) + \delta X^\mu(y) \tag{A.97}$$

where $\delta X^\mu(y)$ is an arbitrary function of y that vanishes at the endpoints of the integration region of the integral. The action is:

$$I = \int \mathscr{L}_S d^4y \tag{A.98}$$

We will show that a variation in $X^\mu(y)$ leads to a conserved energy-momentum tensor. But we will use integrals of the Lagrangian density since it provides a simpler derivation of the result. Under the variation of eq. A.97 we find

$$\delta\phi = \phi(X(y) + \delta X^\mu(y)) - \phi(X(y))$$

$$= \delta X^\mu\ \partial\phi/\partial X^\mu \tag{A.99a}$$

$$\delta(\partial\phi/\partial X^\nu) = \delta X^\mu\ \partial(\partial\phi/\partial X^\mu)/\partial X^\nu \tag{A.99b}$$

$$\delta(\partial X^\mu/\partial y^\nu) = \partial(\delta X^\mu)/\partial y^\nu \tag{A.99c}$$

The integral in eq. A.98 changes by

$$\delta I = \int d^4y\, \delta\mathscr{L}_s = \int d^4y\, [\, \delta(J\mathscr{L}_F) + \delta\mathscr{L}_C\,] \qquad \text{(A.100a)}$$

which becomes:

$$\delta I = \int d^4y\, [\delta X^\mu\, \partial(J\mathscr{L}_F)/\partial X^\mu + \partial(\delta X^\mu \partial\mathscr{L}_C\, / \partial(\partial X^\mu/\partial y^\nu))/\partial y^\nu] \quad \text{(A.100b)}$$

due to the equations of motion of $X^\mu$ (eq. A.19b) in $X^\mu$s role. Since the second term is a total divergence its contribution to $\delta I$ is zero. Thus we can express eq. A.100b as:

$$\delta I = \int d^4y\, [\, J\, \delta\mathscr{L}_F + \mathscr{L}_F\, \delta J\,] \qquad \text{(A.101)}$$

realizing that the Jacobian J depends on y and thus X:

$$\delta J = \delta X^\mu\, \partial J/\partial X^\mu \qquad \text{(A.102)}$$

A partial integration gives

$$\mathscr{L}_F\, \delta J = \delta X^\mu\, \partial(\, J\mathscr{L}_F)/\partial X^\mu - \delta X^\mu\, J\, \partial\mathscr{L}_F/\partial X^\mu \qquad \text{(A.103)}$$

Evaluating $\delta\mathscr{L}_F$ we find:

$$\delta\mathscr{L}_F = \partial\mathscr{L}_F / \partial\phi\, \delta\phi + \partial\mathscr{L}_F\, / \partial(\partial\phi/\partial X^\mu)\, \delta(\partial\phi/\partial X^\mu) \qquad \text{(A.104)}$$

which gives

$$\delta\mathscr{L}_F = \delta X^\nu\, \partial/\partial X^\mu\, [\partial\mathscr{L}_F / \partial(\partial\phi/\partial X^\mu)\, \partial\phi/\partial X^\nu\,] \qquad \text{(A.105)}$$

using the equations of motion eq. A.31, and using eq. A.99b. Combining eqs. A.100, A.101, A.103 and A.105 we obtain:

$$\int d^4y\, J\, \delta X^\nu\, \partial/\partial X_\mu\, \mathscr{T}_{F\mu\nu} = \int d^4X\, \delta X^\nu\, \partial/\partial X_\mu\, \mathscr{T}_{F\mu\nu} = 0 \qquad \text{(A.106)}$$

where

$$\mathscr{T}_{F\mu\nu} = -\, g_{\mu\nu}\, \mathscr{L}_F + \partial\mathscr{L}_F / \partial(\partial\phi/\partial X_\mu)\, \partial\phi/\partial X^\nu \qquad \text{(A.107)}$$

after some manipulations. Since $\delta X^{\nu}$ is an arbitrary function of y the differential conservation law follows:

$$\partial/\partial X_{\mu}\,\mathscr{T}_{F\mu\nu} = 0 \tag{A.108}$$

Eq. A.108 implies the energy-momentum vector

$$P_{F\beta} = \int d^3X\,\mathscr{T}_{F0\beta} \tag{A.109}$$

is conserved:

$$\partial/\partial X^0\,P_{F\beta} = 0 \tag{A.110}$$

The hamiltonian density (eq. A.83) is

$$\mathscr{H}_F = \mathscr{T}_{F0\beta} \tag{A.111}$$

Thus the field energy

$$H_F = P_{F0} = \int d^3X\,\mathscr{T}_{F00} \tag{A.112}$$

is conserved with respect to the "time" $X^0$. Later we will see that $H_F$ is trivially conserved in the Coulomb gauge of $X_{\mu}$. (We will also establish an electromagnetic-like quantum field theory for $X_{\mu}$ with gauge invariance.) In other gauges the conservation of $H_F$ is not trivial.

## Separable Lagrangians and Angular Momentum Conservation

We can also verify Lorentz invariance and obtain the form of the conserved angular momentum for a separable Lagrangian by considering the effect of an infinitesimal Lorentz transformation. We will consider the case of a scalar field $\phi$.

Under an infinitesimal Lorentz transformation as specified by eqs. A.60a – A.60d the separable Lagrangian changes by

$$\delta\mathscr{L}_s = \in^{\mu\nu} y_{\nu}\,\partial\mathscr{L}_s/\partial y^{\mu} \tag{A.113}$$

which can also be expressed as

$$\delta\mathcal{L}_s = \partial\mathcal{L}_s/\partial\phi\,\delta\phi + \partial\mathcal{L}_s/\partial(\partial\phi/\partial X^\mu)\,\delta(\partial\phi/\partial X^\mu) + \partial\mathcal{L}_s/\partial X^\mu\,\delta X^\mu + $$
$$+ [\partial\mathcal{L}_s/\partial(\partial X^\mu/\partial y^\nu)]\,\delta(\partial X^\mu/\partial y^\nu) \quad \text{(A.114)}$$

Combining eqs. A.113 and A.114 leads to:

$$\in_{\mu\nu}\partial/\partial y^\sigma \mathcal{M}_s{}^{\sigma\mu\nu} = 0 \quad \text{(A.115)}$$

where

$$\mathcal{M}_s{}^{\sigma\mu\nu} = \mathrm{J}\,\mathcal{M}_F{}^{\sigma\mu\nu} + \mathcal{M}_C{}^{\sigma\mu\nu} + \mathcal{M}_M{}^{\sigma\mu\nu} \quad \text{(A.116)}$$

$$\mathcal{M}_F{}^{\sigma\mu\nu} = (g^{\mu\sigma}y^\nu - g^{\nu\sigma}y^\mu)\mathcal{L}_F + \partial\mathcal{L}_F/\partial(\partial\phi/\partial y_\sigma)\,(y^\mu\partial\phi/\partial y_\nu - y^\nu\partial\phi/\partial y_\mu) \quad \text{(A.117)}$$

$$\mathcal{M}_C{}^{\sigma\mu\nu} = (g^{\mu\sigma}y^\nu - g^{\nu\sigma}y^\mu)\mathcal{L}_C + $$
$$+ \partial\mathcal{L}_C/\partial(\partial X^\delta/\partial y^\sigma)(g^{\delta\nu}X^\mu - g^{\delta\mu}X^\nu + y^\mu\,\partial X^\delta/\partial y_\nu - y^\nu\,\partial X^\delta/\partial y_\mu) \quad \text{(A.118)}$$

$$\mathcal{M}_M{}^{\sigma\mu\nu} = \mathcal{L}_F\partial\mathrm{J}/\partial(\partial X^\delta/\partial y^\sigma)(g^{\delta\nu}X^\mu - g^{\delta\mu}X^\nu + y^\mu\,\partial X^\delta/\partial y_\nu - y^\nu\,\partial X^\delta/\partial y_\mu) \quad \text{(A.119)}$$

where the third term originates in the dependence of J on derivatives of $X^\mu$. Eq. A.117 was obtained in part by using the identity:

$$\partial\mathcal{L}/\partial(\partial\phi/\partial y^\sigma) = \partial\mathcal{L}/\partial(\partial\phi/\partial X^\alpha)\,\partial y^\sigma/\partial X^\alpha \quad \text{(A.120)}$$

where $\mathcal{L}$ and $\phi$ have the form specified in eq. A.15.

The conserved angular momentum is:

$$\mathrm{M}_s{}^{\mu\nu} = \int dy\,\mathcal{M}_s{}^{0\mu\nu} \quad \text{(A.121)}$$

with

$$\partial \mathrm{M}_s{}^{\mu\nu}/\partial y^0 = 0 \quad \text{(A.122)}$$

*Angular Momentum and $\mathcal{L}_F$*

An alternate conserved angular momentum can be obtained by considering the "field" part of the Lagrangian $\mathcal{L}_F$ under an infinitesimal Lorentz transformation ($\epsilon_{\mu\nu} = -\epsilon_{\nu\mu}$):

$$X'_\mu = X_\mu + \delta X_\mu \tag{A.123a}$$

$$\delta\phi = \phi(X'(y)) - \phi(X(y))$$

$$= \delta X^\mu\, \partial\phi/\partial X^\mu \tag{A.123b}$$

$$\delta X^\mu = S^\mu{}_a X^a(y) - X^\mu(y) \tag{A.123c}$$

$$= \epsilon^\mu{}_a X^a(y) \tag{A.123d}$$

where $S^\mu{}_a$ is the Lorentz transformation matrix for a vector. (If $X^\mu$ is a gauge field then an additional operator gauge term would have to be added to eq. A.123d.)

The Lagrangian changes by

$$\delta\mathcal{L}_F = \epsilon^{\mu\nu} X_\nu\, \partial\mathcal{L}_F / \partial X^\mu \tag{A.124}$$

under an infinitesimal Lorentz transformation. The change can also be expressed as:

$$\delta\mathcal{L}_F = \partial\mathcal{L}_F / \partial\phi\, \delta\phi + \partial\mathcal{L}_F / \partial(\partial\phi/\partial X^\mu)\, \delta(\partial\phi/\partial X^\mu) \tag{A.125}$$

Combining eqs. A.124 and A.125 leads to:

$$\epsilon_{\mu\nu} \partial/\partial X^\sigma\, \mathcal{M}_{FX}{}^{\sigma\mu\nu} = 0 \tag{A.126}$$

where

$$\mathcal{M}_{FX}{}^{\sigma\mu\nu} = (g^{\mu\sigma}X^\nu - g^{\nu\sigma}X^\mu)\mathcal{L}_F + \partial\mathcal{L}_F/\partial(\partial\phi/\partial X^\sigma)\,(X^\mu\partial\phi/\partial X_\nu - X^\nu\partial\phi/\partial X_\mu) \tag{A.127}$$

The conserved angular momentum associated with the X coordinates is:

$$M_{FX}{}^{\mu\nu} = \int d^3X\, \mathcal{M}_{FX}{}^{0\mu\nu} \tag{A.128}$$

with

$$\partial M_{FX}{}^{\mu\nu}/\partial X^0 = 0 \tag{A.129}$$

The angular momentum density can be written in the familiar form:

$$\mathscr{M}_{FX}{}^{\sigma\mu\nu} = X^\mu \mathscr{T}_F^{\sigma\nu} - X^\nu \mathscr{T}_F^{\sigma\mu} \tag{A.130}$$

using eq. A.107.

## Separable Lagrangians and Internal Symmetries

We will now consider the case of a set of scalar fields $\phi_r$ in a separable Lagrangian with an internal symmetry under a local transformation

$$\phi_r(X) \to \phi_r(X) - i\epsilon\lambda_{rs}\,\phi_s(X) \tag{A.131}$$

If the Lagrangian is invariant under this transformation, then

$$\delta\mathscr{L}_S \equiv \delta\mathscr{L}_F = 0 = \partial\mathscr{L}_F/\partial\phi_r\,\delta\phi_r + \partial\mathscr{L}_F/\partial(\partial\phi_r/\partial X^\alpha)\,\delta(\partial\phi_r/\partial X^\alpha) \tag{A.132}$$

Using the equation of motion eq. A.31, which is satisfied by all components $\phi_r$, we obtain a conserved current:

$$\mathscr{J}^\nu = -i\,\partial\mathscr{L}_F/\partial(\partial\phi_r/\partial X^\nu)\,\lambda_{rs}\,\phi_s \tag{A.133}$$

satisfying

$$\partial\mathscr{J}^\nu/\partial X^\nu = 0 \tag{A.134}$$

The conserved charge is

$$Q = \int d^3X\,\mathscr{J}^0 \tag{A.135}$$

$$\partial Q/\partial X^0 = 0 \tag{A.136}$$

We note eq. A.71 provides a corresponding conservation law for the y coordinate system.

# Appendix B. Invariance of Two-Tier Quantum Field Theory under Special Relativity

## Invariance of Two-tier QFT under Special Relativistic Transformations

Turning now to the issue of invariance under special relativistic transformations we begin by noting that the transverse gauge of the $Y^a$ field is not manifestly relativistic. In addition two-tier Feynman propagators calculated in the transverse gauge are also not manifestly relativistic.

The situation is similar to the case of the electromagnetic field yet differs because the $Y^a$ field plays the role in coordinates in the lagrangian. We will now show that the manifestly Lorentz invariant two-tier *Lorentz gauge formulation* is equivalent to the transverse gauge formulation. The classical Lorentz gauge condition:

$$\partial Y^\mu/\partial y^\mu = 0 \tag{B.1}$$

is too stringent to make into an operator relation. We will therefore define the Lorentz gauge formulation of the $Y^a$ quantization by implementing a condition on the space of physical states.

We begin with the covariant equal time commutation relations:

$$[Y^\mu(\mathbf{y}, y^0), Y^\nu(\mathbf{y}', y^0)] = [\pi^\mu(\mathbf{y}, y^0), \pi^\nu(\mathbf{y}', y^0)] = 0 \tag{B.2}$$

$$[\pi^\mu(\mathbf{y}, y^0), Y^\nu(\mathbf{y}', y^0)] = i\eta^{\mu\nu}\delta^3(\mathbf{y} - \mathbf{y}') \tag{B.3}$$

where

$$\pi^\mu = \partial\mathscr{L}_C/\partial Y_\mu' \tag{B.4}$$

$$Y^{\mu\prime} = \partial Y^\mu/\partial y^0 \tag{B.5}$$

The Fourier expansion of $Y^\mu$ is:

$$Y^\mu(y) = \int d^3k\, N_0(k)[\, a^\mu(k)\, e^{-ik\cdot y} + a^{\mu\dagger}(k)\, e^{ik\cdot y}] \tag{B.6}$$

where

$$N_0(k) = [(2\pi)^3 2\omega_k]^{-1/2} \tag{B.7}$$

and

$$\omega_k = (\mathbf{k}^2)^{1/2} = k^0 \tag{B.8}$$

with $k^\mu k_\mu = 0$.

The commutation relations of the Fourier coefficient operators are:

$$[a^\mu(k), a^{\nu\dagger}(k')] = -\eta^{\mu\nu}\delta^3(\mathbf{k} - \mathbf{k}') \tag{B.9}$$

$$[a^{\mu\dagger}(k), a^{\nu\dagger}(k')] = [a^\mu(k), a^\nu(k')] = 0 \tag{B.10}$$

It will be convenient to divide the Y field into positive and negative frequency parts:

$$Y^{\mu+}(y) = \int d^3k\, N_0(k) a^\mu(k)\, e^{-ik\cdot y} \tag{B.11}$$

and

$$Y^{\mu-}(y) = \int d^3k\, N_0(k)\, a^{\mu\dagger}(k)\, e^{ik\cdot y} \tag{B.12}$$

We define

$$a^4(k) = i\, a^0(k) \tag{B.13}$$

and

$$a^{4\dagger}(k) = i\, a^{0\dagger}(k) \tag{B.14}$$

with the resulting commutation relations:

$$[a^\mu(k), a^{\nu\dagger}(k')] = \delta^{\mu\nu}\delta^3(\mathbf{k} - \mathbf{k}') \tag{B.15}$$

for $\mu,\nu = 1, 2, 3, 4$.

Having redefined the operators we then follow the familiar Gupta-Bleuler procedure and introduce an indefinite Dirac metric $\eta$ with $\eta^2 = 1$ and $\eta^\dagger = \eta$ that will enable us to avoid negative probabilities when inner products are calculated. In particular we note,

$$\eta a^4(k) = -a^4(k)\eta \tag{B.16}$$

$$\eta a^i(k) = a^i(k)\eta \tag{B.17}$$

for $i = 1, 2, 3$. Thus

$$\eta = (-1)^{n_4} \tag{B.18}$$

where $n_4$ is the number of time-like $Y^4$ "particles". Let us now define a particle state $\Phi_{n_4}$ with $n_4$ time-like $Y^4$ "particles." Then the time-like raising and lowering operators change the number of particles in a state:

$$a^4(k)\ \Phi_{n_4+1} = (n_4 + 1)^{1/2}\ \Phi_{n_4} \tag{B.19a}$$

and

$$a^{4\dagger}(k)\ \Phi_{n_4} = (n_4 + 1)^{1/2}\ \Phi_{n_4+1} \tag{B.19b}$$

We now chose the coordinate system so the z direction is the direction of "propagation." (A general specification of the coordinate system does not change the result.) Consequently, with the **k** direction fixed, we can write

$$\partial Y^1/\partial y^1 + \partial Y^2/\partial y^2 = 0 \tag{B.20}$$

and the Lorentz gauge condition, which becomes a condition on the physical states,[55] denoted $\Phi_L$, can be written as:

$$(a^3(k) + i\ a^4(k))\Phi_L = 0 \tag{B.21}$$

for all k.

Any physical state $\Phi_L$ can be written as a superposition of states with sharp $Y^3$ and $Y^4$ particle numbers $\Phi_{n,m}$:

$$\Phi_{Ln} = \sum^{n} c_m\ \Phi_{n,n-m} \tag{B.22}$$

where the number of transverse Y "particles" is fixed. In fact, due to our unitarity requirement the number of transverse Y "particles" is always fixed to zero in physical states.

Defining the Lorentz operator

$$L = \partial Y^\mu/\partial y^\mu = L_+ + L_- \tag{B.23}$$

where we separate L into positive and negative frequency parts as in eqs. B.11 and B.12 we see that the Lorentz gauge condition (eq. B.21) becomes

$$L_+\Phi_L = 0 \tag{B.24}$$

---

55 Similar approaches to defining the set of physical states appear in string and superstring theories. See, for example, Bailin and Love ( 1994) pp. 160 –167 and pp. 186 – 190.

with hermitean conjugate (We reserve the $^\dagger$ superscript for later use: $\Phi_L{}^\dagger = \Phi_L{}^*\eta$.)

$$\Phi_L{}^* L_+{}^\dagger = \Phi_L{}^* \eta L_- = 0 \qquad \text{(B.25)}$$

which is obtained by multiplying the hermitean conjugate of eq. B.24 on the right by $\eta$ and using $L_+{}^\dagger = \eta L_- \eta$ and $\eta^2 = 1$. Therefore the inner product equals zero

$$(\Phi_L{}^* \eta L \Phi_L) = 0 \qquad \text{(B.26)}$$

as does

$$(\Phi_L{}^* \eta L^2 \Phi_L) = 0 \qquad \text{(B.27)}$$

which shows the square of the Lorentz condition also vanishes between physical states.

Now, the consideration of the relation between matrix elements in the covariant Lorentz gauge and the transverse gauge requires us to inquire more deeply as to the nature of the physical states defined above. We note that it is easy to show

$$(\Phi_{Ln}{}^* \eta \Phi_{Lm}) = 0 \qquad \text{(B.28)}$$

if n or m is not equal to zero. In the case n = m = 0 the normalization can be chosen such that

$$(\Phi_{L0}{}^* \eta \Phi_{L0}) = 1 \qquad \text{(B.29)}$$

Thus the states $\Phi_{Lm}$ are zero norm states for m ≠ 0 and orthogonal to the vacuum state $\Phi_{L0}$ which happens to be the vacuum state $|0>$ used in our discussions in the earlier chapters. The state $|0>$ contains no transverse, longitudinal or time-like Y "particles."

The zero norm states containing superpositions of longitudinal and time-like Y "particles" are needed due to their role in supporting the gauge invariance of the theory.

*Demonstration that the Expectation Values of Y Field Products in the Lorentz Gauge Equal their Value in the Transverse Gauge*

We will now show that

$$(\Phi_L{}^* \eta B \Phi_L) = (\Phi_{tr}{}^* \eta B \Phi_{tr}) = (\Phi_{tr}{}^* B_{tr} \Phi_{tr}) \qquad \text{(B.30)}$$

$$(\Phi_{L0}{}^* \eta B \Phi_{L0}) = (\Phi_{tr0}{}^* \eta B \Phi_{tr0}) = (\Phi_{tr0}{}^* B_{tr} \Phi_{tr0}) \equiv <0|B_{tr}|0> \qquad \text{(B.31)}$$

where |0> and <0| are the vacua used in previous sections, where $\Phi_{L0}$ and $\Phi_{tr0} = |0>$ are the Lorentz gauge and transverse gauge vacuum states respectively, and where

$$B = \sum_{\lambda} C(\partial/\partial y_1{}^{\mu}, \partial/\partial y_2{}^{\nu}, \ldots, \partial/\partial y_m{}^{\varrho}) Y_{\lambda_1}(y_1)\, Y_{\lambda_2}(y_2) \ldots Y_{\lambda_n}(y_n) \qquad \text{(B.32)}$$

and

$$B_{tr} = \sum_{\lambda} C(\partial/\partial y_1{}^{\mu}, \partial/\partial y_2{}^{\nu}, \ldots, \partial/\partial y_m{}^{\varrho}) Y_{\lambda_1}{}^{tr}(y_1)\, Y_{\lambda_2}{}^{tr}(y_2) \ldots Y_{\lambda_n}{}^{tr}(y_n) \qquad \text{(B.33)}$$

with C a polynomial.

We now write eqs. B.11 and B.12 in the form

$$Y^{\mu+}(y) = \int d^3k\, N_0(k) e^{-ik\cdot y} \{\sum_{\lambda=1}^{2} \varepsilon^{\mu}(k, \lambda)\, a(k,\lambda) + (k^{\mu}/|\mathbf{k}| - \varepsilon_0{}^{\mu}) a_3(k) + $$
$$+ \varepsilon_0{}^{\mu}[a_3(k) - a_0(k)]\} \qquad \text{(B.34)}$$

and $Y^{\mu-}(y)$ in a corresponding form. Then

$$Y^{\mu+}(y)\Phi_L = \{Y^{\mu+tr}(y) + \partial\Lambda/\partial y_{\mu} + \eta^{\mu 0} D^{+}(y)\}\Phi_L \qquad \text{(B.35)}$$

where $\partial\Lambda/\partial y_{\mu}$ can be eliminated with a gauge transformation and where $D^{+}(y)\Phi_L = 0$. As a result

$$Y^{\mu+}(y)\Phi_L = Y^{\mu+tr}(y)\Phi_L \quad \text{and} \quad \Phi_L{}^{\dagger} Y^{\mu-}(y) = \Phi_L{}^{\dagger} Y^{\mu-tr}(y) \qquad \text{(B.36)}$$

where $\Phi_L{}^{\dagger} = \Phi_L{}^{*}\eta$. Consequently

$$(\Phi_L{}^{\dagger}\, Y^{\mu}(y)\, \Phi_L) = (\Phi_L{}^{\dagger}\, Y^{\mu\, tr}(y)\, \Phi_L) \qquad \text{(B.37)}$$

for all physical Lorentz gauge states $\Phi_L$. More generally

$$(\Phi_L{}^{\dagger}\, B\, \Phi_L) = (\Phi_{tr}{}^{\dagger}\, B\, \Phi_{tr}) = (\Phi_{tr}{}^{\dagger}\, B_{tr}\, \Phi_{tr}) \equiv <0|B_{tr}|0> \qquad \text{(B.38)}$$

with B and $B_{tr}$ given by eqs. B.32 and B.33, thus proving eqs. B.30 and B.31. The proof of eq. B.38 is directly based on the representation of $Y^{\mu\pm}(y)$ in eq. B.34, and

$$[D^{+}(k), Y^{\mu\pm}(y)] = 0 \qquad \text{(B.39)}$$

for all k and y with $D^+(k) = a_3(k) - a_0(k)$, and

$$[D^+(k), D^-(k')] = 0 \quad (B.40)$$

for all k and k' with $D^-(k) = a_3^\dagger(k) - a_0^\dagger(k)$, and

$$D^-(k)\Phi_{L0} = 0 \quad \text{and} \quad \Phi_{L0}^\dagger D^+(k) = 0 \quad (B.41)$$

where $\Phi_{L0}$ is the Lorentz gauge vacuum state. As a result of these relations the only surviving terms in B in eqs. B.30 and B.31 are the commutators of $Y^{\mu-tr}$ and $Y^{\nu+tr}$ with the consequence

$$(\Phi_L{}^* \eta B\Phi_L) = (\Phi_{tr}{}^* \eta B\Phi_{tr}) = (\Phi_{tr}{}^* B_{tr}\Phi_{tr}) \quad (B.30)$$

Therefore we have shown the expectation value of a product of Y fields in the Lorentz covariant Lorentz gauge is equal to the expectation value of the corresponding product of transverse gauge Y fields, thus establishing the Lorentz invariance of two-tier quantum field theories.

## Lorentz Invariant S Matrix Elements

Having established the Lorentz invariance of two-tier quantum field theories we now describe a calculational procedure that transforms seemingly non-covariant amplitudes into manifestly covariant amplitudes.

Consider an interaction with a certain number of incoming particles and (possibly) a different number of outgoing particles. Assume a calculation of an S matrix element is performed according to the rules in Blaha (2003), and the present work, in the Y transverse gauge and in the center of mass of the incoming particles. The total momentum in the center of mass is $P^\mu = (P^0, \mathbf{0})$. The S matrix amplitude thus calculated will have a non-manifestly covariant form in the center of mass frame:

$$S_{ab} = S_{ab}(p^0{}_1, p^0{}_2, \ldots, p^0{}_n, |\mathbf{p}_1|, |\mathbf{p}_2|, \ldots |\mathbf{p}_n|, \mathbf{p}_1\cdot\mathbf{p}_2, \ldots \mathbf{p}_i\cdot\mathbf{p}_j, \ldots) \quad (B.42)$$

It can easily be rewritten in covariant form using the total momentum $P^\mu$:

$$p^0{}_i = P^\mu p_{i\mu} / (P^\nu P_\nu)^{1/2} \quad (B.43)$$

$$|\mathbf{p}_i| = [\,(P^\mu p_{i\mu})^2/(P^\nu P_\nu) - p_i^\mu p_{i\mu}\,]^{1/2} \quad (B.44)$$

$$\mathbf{p}_i \cdot \mathbf{p}_j = (P^\mu p_{i\mu} P^\alpha p_{j\alpha})/(P^\nu P_\nu) - p_i^\mu p_{j\mu} \quad \text{(B.45)}$$

for i, j = 1, 2, …, n. After substitutions are made we obtain a completely covariant form:

$$S_{ab} = S_{ab}(P^\nu P_\nu, P^\alpha p_{1\alpha}, P^\beta p_{2\beta}, \ldots, P^\kappa p_{n\kappa}, p_1^\rho p_{2\rho}, \ldots, p_i^\sigma p_{j\sigma}, \ldots) \quad \text{(B.46)}$$

Lastly we note that a gauge transformation can always be made after a Lorentz transformation in the y coordinates that restores the transverse gauge of the Y field.

## Dirac Metric Operator

The metric operator $\eta$ introduced earlier

$$\eta = (-1)^{n_4} \quad \text{(B.47)}$$

can be combined with the metric operator V (eq. 5.15) introduced in earlier chapters for the transverse gauge

$$V = \exp[-i\pi \sum_{\lambda=1}^{2} \int d^3k\, a^\dagger(k, \lambda) a(k, \lambda)] \quad \text{(B.48)}$$

with the property

$$V\, Y^j(y)\, V^{-1} = -Y^j(y) \quad \text{(B.49)}$$

for j = 1,2,3.

The result is the invariant metric operator for the Lorentz gauge:

$$V_L = \exp[-i\pi \int d^3k\, a^{\mu\dagger}(k) a_\mu(k)] \quad \text{(B.50)}$$

which could be used in inner products such as

$$(\Phi_{L0}{}^* V_L B \Phi_{L0}) = (\Phi_{tr0}{}^* V_L B \Phi_{tr0}) = (\Phi_{tr0}{}^* V B_{tr} \Phi_{tr0}) \equiv \langle 0 | V B_{tr} | 0 \rangle \quad \text{(B.51)}$$

that occur in the evaluation of S matrix elements.

We note the following properties of $V_L$:

$$V_L^\dagger = V_L^{-1} = V_L \quad \text{(B.52)}$$

$$V_L^2 = I \quad \text{(B.53)}$$

# REFERENCES

Akhiezer, N. I., Frink, A. H. (tr), 1962, *The Calculus of Variations* (Blaisdell Publishing, New York, 1962).

Bailin, D. & Love, A., 1994, *Supersymmetric Gauge Field Theory and String Theory* (Institute of Physics Publishing, Philadelphia, PA, 1994).

Bergmann, P. G., 1942, *Introduction to the Theory of General Relativity* (Prentice-Hall, Englewood Cliffs, NJ, 1942).

Bjorken, J. D., Drell, S. D., 1965, *Relativistic Quantum Fields* (McGraw-Hill, New York, 1965).

Blaha, S., 2002, *Cosmos and Consciousness Second Edition* (Pingree-Hill Publishing, Auburn, NH, 2002).

Blaha, S., 2003, *A Finite Unified Quantum Field Theory of the Elementary Particle Standard Model and Quantum Gravity Based on New Quantum Dimensions™ and a New Paradigm in the Calculus of Variations* (Pingree-Hill Publishing, Auburn, NH, 2003).

Blaha, S., 2004, *Quantum Big Bang Cosmology: Complex Space-time General Relativity, Quantum Coordinates, Dodecahedral Universe, Inflation, and New Spin 0, ½, 1 & 2 Tachyons & Imagyons* (Pingree-Hill Publishing, Auburn, NH, 2004).

Blaha, S., 2005, *The Equivalence of Elementary Particle Theories and Computer Languages: Quantum Computers, Turing Machines, Standard Model, Superstring Theory, and a Proof that Gödel's Theorem Implies Nature Must Be Quantum* (Pingree-Hill Publishing, Auburn, NH, 2005).

Bogoliubov N. N., & Shirkov, D. V., Volkoff, G. M. (tr), 1959, *Introduction to the Theory of Quantized Fields* (Wiley-Interscience, New York, 1959).

Buck, R. C., (1956), *Advanced Calculus* (McGraw-Hill Publishing, New York, 1956).

Cottingham, W. N. and Greenwood, D. A., 1998, *An Introduction to the Standard Model of Particle Physics* (Cambridge University press, Cambridge, UK, 1998).

Dicke, R. H., 1970, *Gravitation and the Universe* (American Philosophical Society, Philadelphia, 1970).

Dodelson, S., 2003, *Modern Cosmology* (Academic Press, London, 2003).

Donoghue, J. F., Golowich, E. and Holstein, B. R., 1992, *Dynamics of the Standard Model* (Cambridge University Press, Cambridge, UK, 1992).

Eddington, A. S., 1924, *The Mathematical Theory of Relativity* (Cambridge University Press, Cambridge, UK, 1924).

Eddington, A. S., 1995, *Space, Time & Gravitation* (Cambridge University Press, Cambridge, UK, 1995).

Gelfand, I. M., Fomin, S. V., Silverman, R. A. (tr), 2000, *Calculus of Variations* (Dover Publications, Mineola, NY, 2000).

Giaquinta, M., Modica, G., Souchek, J., 1998, *Cartesian Coordinates in the Calculus of Variations* Volumes I and II (Springer-Verlag, New York, 1998).

Giaquinta, M., Hildebrandt, S., 1996, *Calculus of Variations* Volumes I and II (Springer-Verlag, New York, 1996).

Goertzel, G., Tralli, N., 1960, *Some Mathematical Methods of Physics* (McGraw-Hill, New York, 1960).

Goldstein H., 1965, *Classical Mechanics* (Addison-Wesley, Reading, MA 1965).

Guth, A. H., 1997, *The Inflationary Universe* (Perseus Books, Cambridge, MA, 1997).

Hamermesh, M., 1962, *Group Theory* (Addison-Wesley Publishing, New York, 1962).

Heitler, W., 1954, *The Quantum Theory of Radiation* (Oxford University Press, London, 1954).

Huang, K., 1992, *Quarks, Leptons & Gauge Fields Second Edition* (World Scientific, River Edge, NJ, 1992).

Huang, K., 1998, *Quantum Field Theory* (John Wiley, New York, 1998).

Hübsch, T., 1994, *Calabi-Yau Manifolds* (World Scientific, London, 1994).

Jost, J., Li-Jost, X., 1998, *Calculus of Variations* (Cambridge University Press, New York, 1998).

Kaku, M., 1999, *Introduction to Superstrings and M-Theory Second Edition* (Springer-Verlag, New York, 1999).

Kaku, M., 1993, *Quantum Field Theory* (Oxford University Press, New York, 1993).

Kreyszig, E., 1991, *Differential Geometry* (Dover Publications, New York, 1991).

Landau, L. D. and Lifshitz, E. M., 1962, *The Classical Theory of Fields* (Addison-Wesley, New York, 1962).

Lang, S., 1987, *Introduction to Complex Hyperbolic Spaces* (Springer-Verlag, New York, 1987).

Mach, E., 1991, *Die Mechanik in ihrer Entwicklung* (Wissenschaftliche Buchgesellschaft, Berlin, 1991).

Magnus, W. and Oberhettinger, F., 1949, *Formulas and Theorems for the Special Functions of Mathematical Physics* (Chelsea Publishing Co., New York, 1949).

Misner, C. W., Thorne, K. S., Wheeler, J. A., 1973, *Gravitation* (W. H. Freeman, New York, 1973).

Pauli, W., 1958, *The Theory of Relativity* (Pergamon Press, London, 1958).

Polchinski, J., 1998, *String Theory* (Cambridge University Press, New York, 1998).

Quigg, C., 1997, *Gauge Theories of the Strong, Weak and Electromagnetic Interactions* (Westwood Press, 1997).

Rubin, V. C. and Coyne, G. V., *Large-Scale Motions in the Universe* (Princeton University Press, Princeton, 1988).

Sagan, H., 1993, *Introduction to the Calculus of Variations* (Dover Publications, Mineola, NY, 1993).

Schutz, B. F., 2002, *A First Course in General Relativity* (Cambridge University Press, Cambridge, UK, 2002).

Schwerdtfeger, H., 1979, *Geometry of Complex Numbers: Circle Geometry, Moebius transformations, Non-Euclidean Geometry* (Dover Publications, New York, 1979).

Smolin, L., 2001, *Three Roads to Quantum Gravity* (Basic Books, New York, 2001).

Streater, R. F. and Wightman, 2000, A. S., *PCT, Spin and Statistics, and All That* (Princeton University Press, Princeton, NJ, 2000).

Synge, J. L., 1960, *Relativity, The General Theory* (North-Holland, Amsterdam, 1960).

Weinberg, S., 1995, *The Quantum Theory of Fields Volume I* (Cambridge University Press, New York, 1995).

Weinberg, S., 1996, *The Quantum Theory of Fields Volume II* (Cambridge University Press, New York, 1996).

Weyl, H., 1950, *Space, Time, Matter* (Dover, New York, 1950).

# INDEX

# About the Author

Stephen Blaha is an internationally known physicist with extensive interests in Science, the Arts, and Technology. He received his Ph.D. in Theoretical Physics from Rockefeller University (NY). He has written a highly regarded book on physics, consciousness and philosophy – *Cosmos and Consciousness*, a book on Science and Religion entitled *The Reluctant Prophets*, a book applying physics concepts to the history of civilizations, and books on Java and C++ programming. He developed a mathematical theory of civilizations that is described in *The Life Cycle of Civilizations*. Recently he completed a major new study of Cosmology: *Quantum Big Bang Cosmology: Complex Space-time General Relativity, Quantum Coordinates, Dodecahedral Universe, Inflation, and New Spin 0, ½, 1 & 2 Tachyons & Imagyons*. He has served on the faculties of several major universities. He was an Associate of the Harvard Physics Faculty for twenty years (1983-2003). He was also a Member of the Technical Staff at Bell Laboratories, a member of management at the Boston Globe Newspaper, a Director of Wang Laboratories, and President of Blaha Software Inc and Janus Associates Inc. (NH). Dr. Blaha is noted for contributions to elementary particle theory, mathematics, condensed matter physics and computer science.

Among other achievements he was a co-discoverer of the "r potential" for heavy quark binding developing the first (and still the only demonstrable) non-abelian gauge theory with an "r" potential; first suggested the existence of topological structures in superfluid He-3; first proposed Yang-Mills theories would appear in condensed matter phenomena with non-scalar order parameters; first developed a grammar-based formalism for quantum computers and applied it to elementary particle theories; first developed a new form of quantum field theory without divergences (thus solving a major 60 year old problem that enabled a unified theory of the Standard Model and Quantum Gravity without divergences to be developed); first developed a formulation of complex General Relativity based on analytic continuation from real space-time; first developed a generalized non-homogeneous Robertson-Walker metric that enabled a quantum theory of the Big Bang to be developed without singularities at t = 0; first generalized Cauchy's theorem and Gauss' theorem to complex curved multi-dimensional spaces; first developed a physically acceptable theory of faster-than-light particles – tachyons – of any spin; first showed a universe with three complex spatial dimensions has an icosahedral symmetry; first developed the form of the composition of extrema in the Calculus of Variations; first suggested that inflationary periods in the history of the universe were not needed; first proved Gödel's Theorem implies Nature must be quantum, and first developed a quantitative harmonic oscillator-like model of the life cycle, and interactions, of civilizations.

Blaha was also a pioneer in the development of UNIX for financial and scientific applications, database benchmarking, and networking (1982); in the development of Desktop Publishing (1980's); and developed a hybrid shell programming technique (1982) that was a precursor to the PERL programming language. He received Honorable Mention in the Gravity Research Foundation Essay Competition in 1978, and was nominated for three "Awards for Technical Excellence" in 1987 by PC Magazine for PC software products that he designed and developed. His email address is sblaha777@yahoo.com.

www.ingramcontent.com/pod-product-compliance
Lightning Source LLC
Chambersburg PA
CBHW061415210326
41598CB00035B/6221
* 9 7 8 0 9 7 4 6 9 5 8 3 9 *